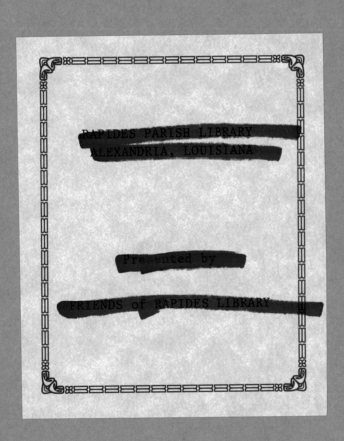

One Universe

At Home in the Cosmos

Rose Center for Earth and Space / American Museum of Natural History

One Universe
At Home in the Cosmos

NEIL DE GRASSE TYSON

CHARLES LIU

ROBERT IRION

JOSEPH HENRY PRESS
WASHINGTON, D.C.

JOSEPH HENRY PRESS
2101 Constitution Avenue, N.W.
Washington, D.C. 20418

The Joseph Henry Press, an imprint of
the National Academy Press, was created
with the goal of making books on science,
technology, and health more widely
available to professionals and the public.
Joseph Henry was one of the founders
of the National Academy of Sciences and
a leader of early American science.

*The authors would like to thank the National
Academy of Sciences, the National Academy
of Engineering, and the Institute of Medicine
for their generous support of this project.*

CONTENTS

Introduction
Our Connection to the Universe

Millions of stars fill the sky near the center of our galaxy, the Milky Way. As we gaze at these dazzling but distant points of light, it's hard to imagine that many of them may be suns in planetary systems like our own.

We live in a universe filled with wonders. Comets hang like celestial torches before fading on their long journeys into space. The sun descends in a golden blaze on a summer evening, and countless stars spill from zenith to horizon through the dark night. At such moments, our cosmos inspires awe.

However, we rarely feel connected to the cosmos. We live at a hectic pace on a warm planet, insulated from the universe by the bright blue dome of sky. At night, when the heavens open up to us, we seldom cast more than a glance overhead. Even when we do notice the grandeur of the universe, it seems utterly separate from our lives. Planets, stars, and galaxies appear so small to our eyes that we cannot comprehend their enormous sizes, so far away that we cannot grasp the vast gulf of space between them, and so exotic that we cannot understand how they work. Our experiences on Earth seem so different from these wonders that nature surely must have followed another set of rules in creating them. Can we ever hope to divine those cosmic principles?

The answer is a resounding "Yes." A deep insight has emerged from astronomy and physics: The basic forces, quantities, and processes that govern our lives on Earth and that govern the workings of the universe are one and the same. In fact nature's laws are fewer in number, and often simpler, than the laws that human societies invent. We can study the natural laws on our planet and in our neighborhood in space, then use those laws to understand the behaviors of objects that lie forever out of reach. In so doing, we have learned that no wall separates our Earth and sky from the rest of the cosmos. We live in One Universe.

Some of those connections are easy to see. A crystal hanging in a window lights the room with bands of color on a sunny day. We use more elaborate crystals to break up light from stars and galaxies. Special instruments extract hidden details from those delicate rainbows, revealing what the objects are made of and how they move through space. Baseball fans watch the cosmos at work when they follow the arc of a home run soaring into the bleachers. The arc is a perfect illustration of the ever-present force of gravity, which pins us to the ground, keeps the Moon in orbit around Earth, and steers our Sun through our Milky Way galaxy. The Moon and the Sun also exert gravitational pulls on Earth, creating tides that we see as the twice-daily ebb and flow of the ocean. Stronger tides elsewhere in the universe turn the insides of moons to mush and stretch pairs of closely orbiting stars into egglike shapes.

Other connections come from watching things spin, a property that applies to nearly everything in space. The whirl of a gyroscope, as children know, prevents it from toppling on its side. Telescopes in space take advantage of that same principle by using three gyroscopes to keep a steady aim. On a larger scale, Earth's daily rotation on its axis stirs our atmosphere and stretches storms into spiral shapes. Other planets display similar stormy patterns, such as Jupiter's Great Red Spot.

Some of our links to the cosmos are more surprising, for they involve events too extreme to occur on Earth. For instance, the largest stars blow up in titanic blasts that seed the galaxy with heavy elements, such as iron, calcium, and silicon. These elements come only from stars; the universe has no other way to create them. They drift into clouds of gas and dust which collapse into a new generation of stars, planets, and—in our case—life. In other words, dying stars forged the elements that compose the blood in our veins, the bones in our bodies, and the chips in our computers. The stuff of stars is all around us even though the stars themselves seem so inaccessible.

Our awareness of these connections has grown as we have studied the natural world for thousands of years. The earliest natural philosophers—Plato, Aristotle, and Archimedes among them—tried to use their five senses, in combination with logic and reason, to explain the cosmos. However, their preconceptions got in the way. Earth sat unmoving at the center of the universe, they believed, and the celestial bodies moved around it in perfect patterns. These beliefs also affected their view of physical principles on Earth. For instance, Aristotle asserted that heavier objects fall faster than light objects, but he never bothered to put that claim to the test.

Our modern approach to gathering knowledge about the universe draws from traditions established by Galileo Galilei, Isaac Newton, Albert Einstein, and other great minds of the past several centuries. These physicists didn't care whether their results conformed to common-sense views about how the universe worked. Rather, they devised careful theories based on repeated experiments and mathematical analysis. Their theories strove to explain some of what was not understood, predict previously unknown phenomena, and consistently confirm their predictions by further tests. Describing nature as it was, not as the scientists supposed it to be, was at the heart of this scientific method.

In this way, for example, Newton assembled methodical descriptions of how objects move through the universe at everyday speeds. Much later, Einstein found more basic rules that explain how all objects move, even those that travel close to the speed of light. Newton's work was still correct, but it became a special part of Einstein's overall theory. This process is typical of science. Modern technology provides more penetrating insights about nature, leading to new theories that are more accurate but increasingly simpler at their cores. Rarely does a completely surprising phenomenon arise that forces us to overturn all aspects of an existing theory.

Today, we benefit from the creative use of technology to extend our vision far beyond Earth's surface and our solar system. Telescopes, spectrographs, electronic cameras, and other tools collect data every night from the farthest corners of the cosmos, revealing what our unaided eyes could never see. We also use computers to simulate processes that we cannot duplicate in laboratories on Earth. For instance, computer models shed light on the pervasive influence of gravity, which extends invisible tendrils across the entire cosmos. The programs calculate billions of years of gravitational interactions among galaxies to show why the universe looks the way it does today.

These scientific pursuits rely upon studies of three fundamental aspects of nature: motion, matter, and energy. Motion is a logical starting point, since everything moves—from the atoms in stationary objects to the most distant galaxies. Ancient observers founded the science of astronomy by charting the motions of the Sun, Moon, stars, and planets in painstaking detail. Today, our telescopes and observing tools are sensitive enough to detect planets around other stars. But we have learned that the motions of celestial objects are ever-changing. Just slight alterations in their paths through space can have dramatic consequences. For that reason we keep a wary eye on space, watching for comets and asteroids that could be headed our way.

Matter comes in many forms, from the familiar objects in our homes to exotic varieties in space. These diverse substances share a list of ingredients: about 100 unique elements. Most are in short supply—our universe consists almost entirely of hydrogen and helium, with just a dash of heavier elements thrown in. On Earth we are accustomed to seeing matter within the narrow range of temperatures and pressures that make life possible. But such conditions are rare elsewhere. Just a few atoms drift here and there in the cold spaces between stars and galaxies. Within a star, it's hot and dense enough to ignite nuclear fusion—an energy bonanza we haven't yet harnessed. The strangest objects in the universe are forms of matter we will never create here: neutron stars and black holes.

When matter is put in motion, it emits energy. Energetic outbursts throughout the cosmos give us insights into objects that we otherwise would never detect. A star explodes somewhere once every second, blasting light and ghostly particles called neutrinos into space. Gas plunges into black holes at the centers of galaxies, releasing waves of x-rays. The Sun is a constantly churning ball of charged gas laced with magnetic fields that writhe and snap, propelling dangerous flares toward Earth. Our eyes are tuned to a tiny part of this rich display of energy, but the rest bombards us and our planet constantly. We have devised clever ways to see those elusive waves, from giant radio receivers on the ground to x-ray and gamma-ray telescopes in orbit.

Beyond these ongoing studies, we face steeper challenges ahead. Some of the questions at the frontiers of cosmological science today seem extraordinarily hard to address: Have matter and energy combined to create life elsewhere? What are the essential ingredients of matter? Does a single theory of physics describe the behaviors of all forces and particles in the universe? What sparked the birth of the universe? What is its ultimate fate, after all the suns have burned out?

We will explore these questions with the same scientific tools that have revealed the universal laws of nature so far. For instance, searches for life on other planets are planned or under way with space probes and observations from Earth. Particle colliders probe ever more deeply into the nesting Russian doll of the atom. The bizarre consequences of modern physics suggest that the tiniest components of matter, which dwell in a Wonderland that we are straining to comprehend, may have sown the seeds of the universe itself. As for the future, we have found hints that an eerie force of repulsion permeates the universe, forcing it to expand more quickly as time goes on.

Our studies of the distant universe move forward because we are confident that the principles of physics governing nature on Earth also apply throughout the cosmos. Basic quantities such as the strength of gravity or the charge of an electron remain the same—within the limits of our abilities to measure them—no matter where one goes. Atoms shine or decay radioactively in a laboratory on Earth in the same way as they do billions of light-years across space. Magnetic fields exist everywhere and affect charged particles in the same way.

What's more, our Sun is an ordinary star, like billions of others in the Milky Way. Our galaxy is much like other spiral galaxies in the universe. It's quite likely that our planet is just one of countless rocky planets orbiting stars at hospitable distances—not too hot, not too cold. Five hundred years ago, Nicolaus Copernicus voiced the notion that there is nothing special about our place in the cosmos or the time in which we live. The Copernican principle still holds sway. It gives us the freedom to apply what we know about Earth, the Sun, and the Milky Way to any other location in the cosmos because we assume the laws of nature here are quite ordinary.

On the largest scales of all, we are finding that the universe looks the same in every direction. Any big chunk of space contains galaxies arrayed in similar patterns as any other big chunk. The faint remnants of heat left over from the explosive origin of the universe are smooth across the entire sky to within one part in 100,000. We refer to this large-scale uniformity of the universe as the cosmological principle. It makes it even more likely that the natural laws on our cosmic city block are the same as those elsewhere.

Indeed, as we tour the cosmos, we will find that the behaviors of the largest and smallest objects spring from the same physical principles. Between these extreme scales lies the universe as we know it: grains of sand, babies, jumbo jets, our planet and its neighbors in space. The physics of this comfortable world offers us a template to understand the mysteries of our One Universe.

Motion
Everything Moves

The concentric tracings of stars wheeling above Delicate Arch in Utah's Arches National Park are the product not of stellar movement but of Earth's rotation on its axis, just one of many types of motion that prevail in the universe.

E verything moves. From the molecules of air around us to distant islands of stars, nothing sits still. We hear of comets crashing into planets, black holes gulping streams of gas, and space itself expanding like some vast balloon. And yet the night sky cloaks these cosmic motions. Apart from the wandering Moon and planets and an occasional meteor, the heavens don't seem to change.

Indeed, our eyes cannot see stars moving relative to each other from night to night, or even from one generation to the next. The constellations look essentially the same as those described by astronomers thousands of years ago. The sky's steady patterns make it easy to understand why our ancestors thought the stars were fixed on a giant black sphere that revolved around Earth. Telescopes also reveal countless galaxies suspended in space beyond our home galaxy, the Milky Way, but they seem motionless as well. How can we reconcile the motion we know with the stillness we see?

We can start by looking at our own world in a different way. Many things on our planet appear locked in place even though they move relentlessly. With patience we can see the hour hand swivel around a clock's face or a new rose open its petals to the Sun. If we could somehow watch for millions of years, we would see continents drift across the globe like slabs of ice on a partly frozen lake. Other things seem to move slowly, but only because they are far away. A plane 35,000 feet overhead appears to pass lazily above the clouds, but we know that it speeds 10 times faster than a car zipping past us on the street. Satellites show a hurricane swirling with a grace that belies the fury of its impact on a shoreline. Time and distance hide the true nature of these motions. We see them only when we learn to transcend those barriers, to look beyond the usual rhythm and scope of our world.

Some of the greatest triumphs of scientific inquiry have come from such leaps in vision. In the mid-1950s, geologists learned to read the history book of motion on Earth by finding ancient magnetic imprints preserved in rocks. The imprints revealed the past travels of continents and the constant birth of new crust at the bottom of the sea. With that discovery our planet became a dynamic world. Earthquakes and volcanoes suddenly made sense as the results of those churnings. In a similar way, astronomers learned to break the barrier of time in the cosmos by using telescopes to expose patterns hidden in the light from distant stars. A man named Edwin Hubble studied those patterns to unveil an astonishing fact: Our entire universe is expanding, and at a phenomenal speed. When we look today at the silent night sky, Hubble's discovery remains as awe inspiring as when he announced it in 1929.

The Expanding UNIVERSE

Hubble had access to the 100-inch telescope on Mount Wilson, California, the biggest in the world at the time. During the 1920s, he and astronomer Vesto Slipher of Lowell Observatory in Arizona measured the speeds at which scores of galaxies move through space. They calculated the speeds by detecting subtle shifts in the colors of light emitted by stars in each galaxy. Such shifts reveal a galaxy's motion toward or away from Earth (page 137). One of the first galaxies Slipher studied was the great nebula in Andromeda, which we can see as a fuzzy patch of light with our unaided eyes. He was shocked to find it streaming toward the Milky Way, closing the gap between the two galaxies by 186 miles every second. Most other galaxies analyzed by Hubble and Slipher displayed the opposite motion, receding from us at even higher rates—up to nearly 700 miles per second.

By itself that research was noteworthy. But Hubble's great contribution was to combine the speed measurements with distances to the galaxies. That critical link was provided by rare giant stars that grow brighter and dimmer in regular cycles every few days. Henrietta Swan Leavitt, an astronomer at the Harvard College Observatory, studied these stars, called Cepheid variables, in a pair of galaxies close to the Milky Way. By 1912 she had demonstrated that the length of a Cepheid's cycle depends only on the star's luminosity, or inherent brightness. Bright Cepheids flicker more slowly than faint ones, Leavitt found. It was as though she had discovered that all 60-watt lightbulbs flicker at a certain rate, while all 100-watt bulbs flicker more slowly. Each Cepheid variable has a characteristic "wattage" that astronomers can deduce by observing how long it takes for the star to pulsate.

The Mount Wilson telescope gave Hubble the light-gathering power he needed to spot Cepheid variables in galaxies much farther away than those studied by Leavitt. The first such star he spotted in a distant galaxy was in 1923, in the great Andromeda nebula itself. He marked the discovery with an excited notation of "VAR!" on a photographic plate. During the next several years, Hubble measured the flicker rates of Cepheids in many other galaxies in painstaking detail. He also noted how bright the Cepheids appeared in the telescope. Putting those two quantities together let Hubble determine the distances to the stars and, by association, to their host galaxies. You would do much the same thing to estimate the distance to a house across a dark meadow by observing the lightbulb outside the front door. Perhaps you know roughly how bright a 100-watt bulb appears from half a mile away. If someone changed the light to a 60-watt bulb, you'd have to be much closer to the house for it to appear as bright to your eyes.

When Hubble combined his data, the results were stunning. Except for the few nearest galaxies, each galaxy's speed through space was directly related to its distance from Earth. If one galaxy was twice as far away as another, it moved away from Earth about twice as fast. Five times farther away meant five times faster, and so on. That's the telltale signature of an explosion. Colorful fragments of fireworks expand into the sky in the same way: Those near the center move slowly, while the flaming bits near the edge blast outward fastest of all. Hubble's work showed beyond doubt that the apparent repose of galaxies against the dark vault of space is an illusion. No longer did the cosmos seem a static place. In honor of this discovery, the orbiting Hubble Space Telescope carries Hubble's name. Appropriately, one of the telescope's primary missions is to nail down the rate at which galaxies are flying apart from one another.

Since Hubble's research, we have gradually realized that galaxies do not literally stream through space at warp speeds. Rather, they are carried along for the ride as space itself grows in all three dimensions. To imagine this, suppose that the galaxies are ladybugs sitting on the surface of a rapidly inflating balloon. Each bug sees its neighbors on all sides moving away from it, even though the bugs all think they are sitting still. No bug is at the "center" of the expansion because the expansion happens everywhere on the surface at once. The fabric of the balloon stretches with time, as does the very fabric of space in the cosmos.

This universal expansion gives us the strongest piece of evidence we have about our origins. If the cosmos is growing, it was smaller yesterday than it is now. One year ago it was smaller still. The farther back in time we go, the smaller the cosmos must have been. By following this line of reasoning to its logical conclusion, we know there must have been a time when the entire universe existed as a tiny point—and that

The Expanding Universe

Like ladybugs on the surface of a rapidly inflating balloon, galaxies in the expanding fabric of space in the cosmos move away from one another in all directions. The notion that the universe is expanding grew from the 1929 discovery by Edwin Hubble that the farther away a galaxy is, the faster it moves away from us.

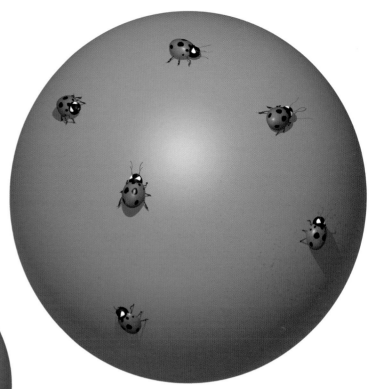
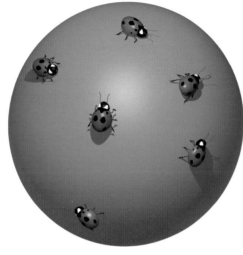

some explosion triggered its rapid growth. As far as we can tell, that cataclysmic event occurred about 13 billion years ago. Its name is now part of our popular culture: the "Big Bang." This mind-boggling scenario raises many questions. However, of all the stories yet devised about the birth of the universe, the Big Bang is best supported by solid scientific evidence.

MOTION Through the Millennia

Hubble's work was a natural climax to thousands of years of curiosity about motion in its many guises. If he had to pick two words to summarize the state of the universe, Hubble might well have chosen those used to open this chapter: "Everything moves." The ancient Greeks understood this idea and captured it in an elegant sentence, translated centuries later into Latin: "Ignorato motu ignorator natura," or "One who knows not motion knows not nature." It's fair to say that to comprehend the universe at large we first must understand how things move in the world around us and in the close confines of our own solar system.

Observations of the Sun, Moon, and planets moving in the sky gave our ancestors their first clues about our place in the cosmos. They watched the Sun arc from east to

A Cosmic Clock

If you look toward the North Star on a clear evening on January 1, you see constellations such as Cassiopeia (*left panel, top*), the Little Dipper, and the Big Dipper. Six hours later (*center panel*), the constellations have rotated one-quarter of the way around the sky, due to Earth's spin on its axis. After 12 hours (*right panel*), our planet has completed half of its daily 24-hour rotation and the constellations have traveled halfway around the sky. Similarly, the constellations appear to circle the North Star over the course of a year as Earth revolves around the Sun. Thus the sky looks the same on July 1 at 6:30 p.m., halfway through the year, as it does on January 2 at 6:30 a.m., halfway through the day. Observations of these and other recurring patterns in the cosmos led to the calendars we use today.

January 1, 6:30 p.m.

west each day. At night the stars moved in the same direction, apparently revolving around a single point in the northern sky. By tracking the positions of the Sun and other stars for thousands of days, these earliest astronomers found that they followed predictable paths with repeating cycles. Skywatchers also monitored the locations of many patterns of stars, or constellations. When certain constellations appeared exactly at sunrise or sunset, the astronomers could determine how long daylight would last, how warm or cold it would get, and how many more days they could count on such weather.

This evolving awareness of the flow of time, patterns in the heavens, and the change of seasons led to the first agricultural societies and civilizations. The stars served as harbingers of many key events in those societies. The cyclical nature of those events led to the creation of the calendar, the system by which we still organize time today. For example, ancient Egyptian farmers knew that when the bright star Sirius rose in the east just ahead of the Sun, it was time for the Nile's annual flood. The warmest weather in the Northern Hemisphere also occurred at the same time. Since Sirius is called the "Dog Star" for its position in the constellation Canis Major (the great dog), we still refer to those warm weeks as the dog days of summer.

Other remnants of early gazing at the sky persist in modern society. The Babylonians divided the annual solar cycle into months, based on a narrow band of 12 constellations through which the Sun traveled during the year. As each star pattern in

January 2, 12:30 a.m.
April 1, 6:30 p.m.

January 2, 6:30 a.m.
July 1, 6:30 p.m.

turn appeared at the sunrise horizon, the Babylonians knew that one-twelfth of the year had passed. Today, this "zodiac" still provides fodder for mildly amusing astrological predictions about our lives on the daily comics pages. Other cultures, notably in China and the Middle East, divided the year into different but equally accurate segments based on the changing shape and position of the Moon. The lunar calendar still dictates the timing of such events as Easter, Passover, Ramadan, and the Chinese New Year.

Ancient astronomers also excelled in their studies of eclipses—alignments with the Sun in which the Moon casts a dark shadow upon Earth or vice versa (page 16). A special cosmic coincidence allows the Moon to blot out the Sun. Although the Sun is 400 times larger than the Moon, it is also 400 times farther away from Earth. The two bodies thus appear almost exactly the same size in the sky, a situation that occurs nowhere else in the solar system. As a result, Earth is the only planet that experiences solar eclipses in such a spectacular fashion. The Moon's shadow sweeps across the landscape at more than 1,000 miles per hour, casting a brief pall of night in the middle of the day. Temperatures drop, birds stop singing, blossoms close, and nature itself seems to pause. An otherworldly glow surrounds the Moon's black disk. That's the corona, the million-degree atmosphere of the Sun that we usually can't see.

Total solar eclipses happen every couple of years, seemingly over random parts of the globe. But observers who tracked the motions of the Moon and the Sun realized

Ancient astronomers also excelled in their studies of eclipses—alignments with the Sun in which the Moon casts a dark shadow upon Earth or vice versa.

that eclipses recur in patterns after an interval called the "Saros cycle," which lasts 18 years, 11 days, and 8 hours. Combining that cycle with Earth's rotation allows astronomers to forecast the exact paths and durations of eclipses far into the future. Such cycles are common in the universe because the simple physical laws that govern motions often lead to repeating patterns.

However, ancient astronomers explained those patterns in a manner very different from the way we explain them today. In their conception of the universe, Earth sat still at the center while everything else rotated around it. This was a perfectly natural assumption. After all, our feet are planted firmly on the ground. We have no sense of whizzing through space or of spinning on an axis at hundreds of miles per hour. In this "geocentric" framework, the Sun, Moon, planets, and stars surrounded Earth on a nested set of spheres. Some of the motions were easy to describe. For example, the stars shone from the outermost shell, which an unseen deity rotated at a steady pace.

Accounting for the motions of the planets required more complicated schemes. Their wanderings included zigzags and loops that simple rotation on invisible spheres could not explain. In the second century A.D., the Greek scientist Ptolemy devised an elaborate model of the heavens in which the planets traveled on small circles, called "epicycles," while also moving around Earth on their larger spheres. This reproduced the curlicues of planetary motion among the stars, although never exactly as astronomers observed them. Even so, Ptolemy's model reigned for nearly 1,500 years in the absence of further advances.

In the sixteenth century, the Polish astronomer Nicolaus Copernicus put forth the next serious model of the solar system, one in which all planets, including Earth, moved around the Sun. (The Greek scientist Aristarchus of Samos is said to have proposed this in the third century B.C., but those writings have not survived.) Copernicus still thought the planets orbited in perfect circles, which isn't quite true. Further, his ideas weren't broadly accepted until long after his death. Even so, they prepared the way for the man considered by many to be the first modern scientist: Galileo Galilei of Italy.

Galileo was the first person to use a telescope for astronomy. In 1610 he found four points of light along a line on both sides of Jupiter. The points seemed to move

Pearly white streamers of the solar corona blaze across the darkened sky during a total solar eclipse. Normally hidden in the Sun's glare, the outer atmosphere of our home star is visible to our eyes only when the Moon blots out the solar disk.

Eclipse Special Effects

Earth is treated to eclipses when the Moon's shadow falls on Earth (solar eclipse) and when Earth casts its shadow on the Moon (lunar eclipse). Total solar eclipses occur because the Moon and the Sun appear almost exactly the same size in the sky. But when the Moon is on the far side of its elliptical orbit, it appears somewhat smaller than the Sun. We then get an annular eclipse in which a bright ring surrounds the Moon.

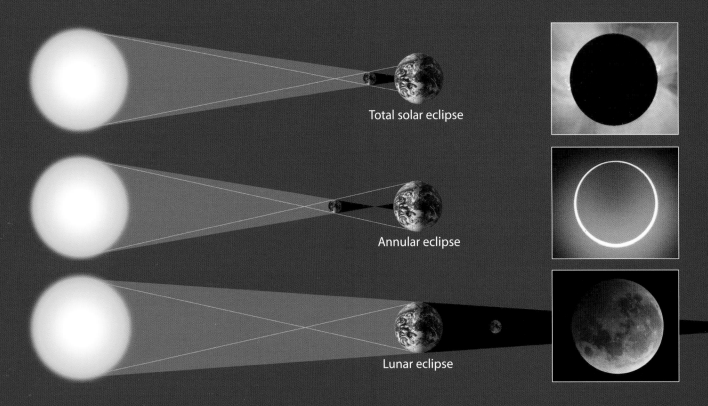

Total solar eclipse

Annular eclipse

Lunar eclipse

Eclipses don't occur every month because the Moon's orbit is tilted about 5 degrees to the plane of Earth's orbit. Only when the Moon and Earth line up at the intersection of the two orbital planes (dashed line) will the Moon's shadow fall on Earth or vice versa. Anywhere else in its orbit, the Moon is either above or below the plane of Earth's orbit, and the shadows of either body fall on empty space.

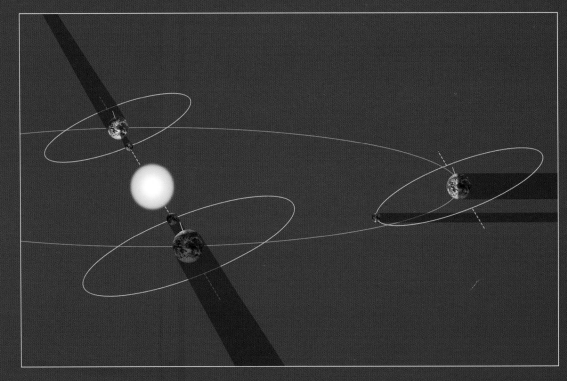

from night to night, but they never strayed far from the planet. Galileo deduced that they were moons orbiting Jupiter, and indeed they were. Today we know them as Io, Europa, Ganymede, and Callisto—the Galilean satellites. He also studied mountains on the Moon and the changing appearance of Venus, which waxes and wanes from crescent to full and back again, just like the Moon. Galileo calculated that if the Sun and Venus both circled Earth, as in Ptolemy's scheme, Venus could never appear as more than a crescent. Thus, its waxing and waning could occur only if Venus circled the Sun rather than Earth.

Galileo's astronomical discoveries, and his stubborn persistence in promoting them, so infuriated the Roman Catholic Church that near the end of his life he was forced to recant his findings. Legend asserts that after his trial, where he was pronounced guilty of heresy, Galileo proclaimed about the planet beneath his feet, "Eppur si muove"—"Nevertheless, it moves." Less disturbing to the church were his investigations into how things move on Earth itself, work that effectively laid the foundation for the science of physics.

Before Galileo's studies, people believed that not only did Earth stand still but that objects upon it preferred to stay at rest. All motion seemed to stop of its own accord, and rather quickly most of the time. Galileo disproved that notion with careful experiments involving metal balls, cylinders, and smooth inclined tracks. His measurements revealed the tendency of a moving object to keep moving until some external force, such as friction, makes it change. We call this concept inertia. An air hockey table illustrates the idea perfectly. Before you plunk quarters into the game, the plastic puck skids just a few inches across the table as friction grinds it to a halt. But when that cushion of air is turned on, almost all friction vanishes. The slightest tap sends the puck gliding toward your opponent's goal without slowing down.

Galileo died in 1642. The English physicist Sir Isaac Newton was born the same year. The world has never witnessed a more symbolic passing of the scientific baton. Newton used his extraordinary intellect to synthesize and extend many of the studies of motion that had preceded him. In his historic book *The Mathematical Principles of Natural Philosophy*, published in 1687, Newton laid out the three laws of motion (page 21) that students still learn today in physics classes.

The Retrograde Mystery

The ancients held that the Sun, Moon, and planets traveled in perfect circles around Earth and inside a sphere of fixed stars. In the second century A.D., the Greek scholar Ptolemy explained inconsistencies in planetary motion—such as the observed retrograde, or backward, motion of Mars— by suggesting that the planets moved in small circles, or epicycles, at the same time that they revolved around Earth (*right*). In the early seventeenth century, the German astronomer Johannes Kepler worked out the real explanation. Because Earth orbits closer to the Sun, it moves more rapidly than does Mars (*below*). As a result, Mars seems to go forward as seen from Earth at positions 1 and 2, then backward when seen at positions 3 and 4, and then forward again when seen at position 5.

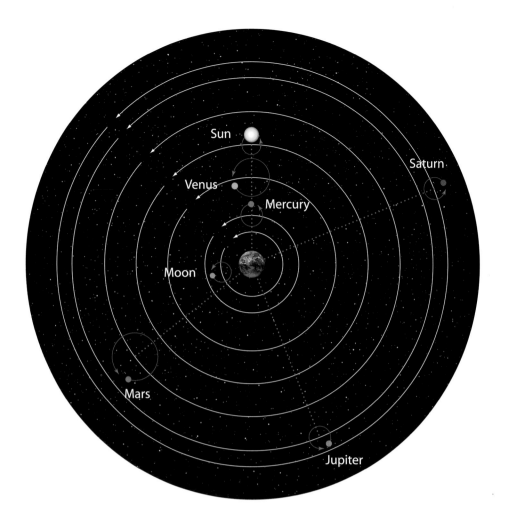

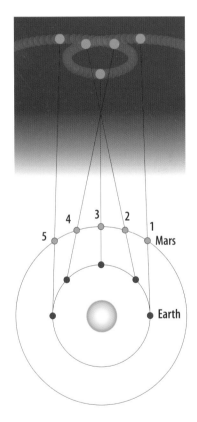

The first law restates and clarifies Galileo's idea of inertia. As Newton described it, an object moving in a straight line at a constant speed will continue with that exact motion until an outside force disturbs it. Since zero speed also qualifies as "traveling" at a constant speed, a corollary to the law of inertia reads: A body at rest will stay that way unless another force acts upon it. We realize that all too well when we have to push a stalled car out of a crowded intersection. In that situation, inertia and friction are powerful hindrances. But in the vast spaces between stars, frictional forces are nearly absent. Thus, spacecraft such as *Voyager 1* and *Voyager 2*, launched in 1977 to explore the outer planets and then escape our solar system, will sail through space for many thousands of years. They will slow down only slightly from the gravitational pull of the star they have left behind.

Newton's second law provides the mathematical recipe for how and why an object can be set into motion. If you apply force to a mass, its velocity changes. That is, it accelerates or decelerates. As the mass of the object goes up, the resulting acceleration goes down in equal measure if the same force is applied. To picture this, imagine standing behind two people wearing roller skates. One is a 90-pound ballerina, and the other is a sumo wrestler who weighs five times as much. If you push on each

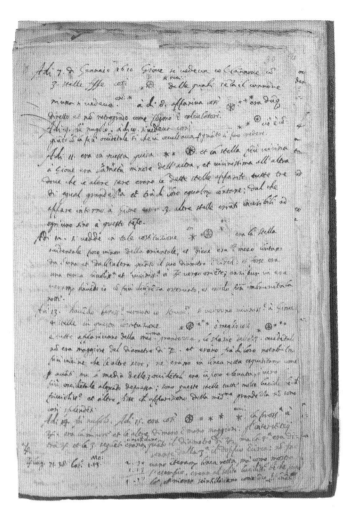

Galileo and Jupiter's Moons

In 1610 Galileo Galilei made meticulous notes of his observations of four wandering "stars" orbiting in a plane around Jupiter. A page from his notebooks shows that he used asterisks to represent the numbers of bodies he had discerned and that he roughly approximated their relative distances from the planet.

After tracking their movements, Galileo realized that the stars were actually moons—now known as the Galilean satellites Io, Europa, Ganymede, and Callisto. As seen in the modern photograph below, Callisto (*far right*) orbits about four times farther from Jupiter than does Io (*near right*), the nearest Galilean moon.

person with equal force (and tact), you will accelerate the ballerina five times more quickly. That ratio holds true in space as well. Astronauts must remember it every time they grapple with objects floating around. It doesn't take much force to stop a wrench passed from one astronaut to another. But if it's a one-ton satellite, it takes just as much effort to control it in space as it does to push your car in neutral on the street.

Newton's third law is justly famous. Paraphrased, we know it as follows: For every action there exists an equal and opposite reaction. This is an easy law to see and understand. If the action is air gushing from the narrow opening of a balloon, the reaction is the balloon jetting crazily around the room. Igniting a rocket's engines propels it off the ground into space. Jumping from a canoe to a dock pushes the canoe out into the water. Even if you agree that a reaction always happens, it's a mental challenge to realize that Earth is the same as that canoe. When you jump into the air, the planet moves in the opposite direction. But because Earth is so much more massive, its response to your leap is minuscule indeed.

Newton's laws hold true for nearly every type of motion on our world as well as throughout the universe. Taken together, they express one of the most crucial physical principles that govern motion in the cosmos: the conservation of momentum.

We calculate an object's momentum by multiplying its mass by how fast (and in what direction) it moves. Momentum cannot be created or destroyed, whether for a single object or a system of objects that interact with one another, unless another force comes into play. We can apply this tenet equally well to understanding the physics of billiards, collisions between asteroids, or the motions of a hundred billion stars in our Milky Way. Without force the total momentum of any system never goes up or down. Once begun, motion continues forever. This is the reason constant motion is the natural state of anything in the universe. This is the reason everything moves.

But is everything in the universe really moving? The newspaper on your coffee table, the tree in your yard, or the building you live in—don't they all stand still around you as you sit reading this book? In fact, they are still—to you. It's all relative.

Galileo and Newton both realized that the measurement of motion depends completely on your frame of reference. Suppose you see a unicyclist ride past at a certain speed. To him you move backward at that same speed, even though you think you're standing still on the sidewalk. If he is juggling at the same time, he sees the balls bob straight up and down from his hands. However, you see the balls move along forward arcs in space as he pedals past. Both viewpoints, or "frames of reference," are equally valid.

Similarly, let's say two cars approach each other on the road, each moving 50 miles per hour. If you stand on the sidewalk, in Earth's frame of reference, you see each car doing 50. But the driver of each vehicle sees the other car zoom toward him at 100 miles per hour. Velocities simply add together in the world of classical relativity as elucidated by Galileo and Newton. That's all well and good unless the cars somehow accelerate to millions of miles per hour. Then, this kind of relativity would fail. When we deal with the realm of superhigh speed, relativity takes on a special form.

In the late nineteenth century, Newton's laws of motion began to break down for objects that move very fast. The American physicists Albert Michelson and Edward Morley tried to add the speed of Earth's revolution around the Sun to the speed of a light beam using a sensitive light-measuring device called an interferometer. They were searching for signs of the "ether," an invisible and unmoving substance believed by physicists of the day to pervade the universe and carry waves of light. To their great surprise, the combined speed of Earth's motion and the ray of light was always exactly the same as that of light alone. Light did not seem to follow the known rules of Newtonian motion.

This puzzle lasted for nearly two decades. Then, in 1905 a theory that explained the startling result arose from the mind of 26-year-old Albert Einstein, a German physicist who worked by day as a patent officer in Switzerland. His mathematical treatise, innocently titled "On the Electrodynamics of Moving Bodies," presented a revolutionary idea that would become known as the special theory of relativity. Einstein asserted that the speed of light—186,282 miles per second—remains constant and can never be exceeded. Further, he said, speed is independent of how quickly an observer might move. Passengers on a spaceship traveling at 186,281 miles per second would still measure their headlight beams streaming away at the full speed of light. Observers on the ground would see the beams moving at exactly the same speed.

Newton's Laws of Motion

Newton's laws, expressed in his words below, contain a profound implication. If an object starts moving, stops moving, speeds up, slows down, or changes its direction, an external force must be at work. As the basis for classical mechanics, Newton's work set the stage for applying these principles to the motions of all objects in the universe.

Law 1 Every body continues in its state of rest, or of uniform motion in a right line, unless it is compelled to change that state by forces impressed upon it.

Law 2 The change of motion is proportional to the motive force impressed and is made in the direction of the right line in which that force is impressed.

Law 3 To every action there is always opposed an equal reaction: or the mutual actions of two bodies upon each other are always equal and directed to contrary parts.

Einstein asserted that the speed of light—186,282 miles per second—remains constant and can never be exceeded.

This theory has peculiar consequences. For example, if that unicyclist could pedal past you at 87 percent the speed of light, you would see him and his unicycle shrink to half their usual lengths. You would see his clock run at half the speed as your own. Further, his mass would be twice as large as when he stands still next to you. These bizarre relativistic effects are important in particle colliders, where physicists boost electrons and protons near the speed of light. They also dictate the behaviors of quasars and other superenergetic objects in the universe, which can expel matter at extremely high speeds.

On Earth special relativity rarely comes into play. Newton's laws suffice for most of our day-to-day activities. Among the exceptions are satellites that use exquisitely accurate clocks, such as those in the Global Positioning System. These satellites orbit the planet at thousands of miles per hour. Their clocks slow down a tiny bit relative to those on the ground, just enough to ruin the precision of their measurements if we ignore the effect.

The constant speed of light has another profound implication for our studies of the universe. On Earth we don't notice a delay as light travels from an object to our eyes. After all, a light beam can dash around the globe nearly eight times a second. But light's fixed speed begins to matter away from our planet. Light from the Sun takes about eight minutes to get here. We never see the Sun as it is now, only as it was eight minutes ago. Similarly, we see the next nearest star, Proxima Centauri, as it was 4.1 years ago—the amount of time its light takes to reach us. We see the nearest big galaxy, Andromeda, as it existed more than 2 million years ago. The light from ever more distant galaxies takes billions of years to reach us, from a time when the universe itself was still young. Indeed, our giant telescopes are the time machines into the past that Jules Verne dreamed about—with the disappointing exception that we cannot physically travel through those portals.

Newton's laws also cease to operate reliably at submicroscopic scales. In the world of Galileo and Newton, one object orbiting another behaves much like a cat on a ramp. It can walk up or down the ramp, pausing at any level it likes. But when physicists early in this century scrutinized the behaviors of atoms, they realized that only certain motions and energies are allowed. All others are forbidden. In this world, an electron orbiting the nucleus of an atom behaves more like a kangaroo on a

The Relativity of Time, Mass, and Length

Most aspects of the physical world can be described in terms of three quantities that we would normally consider easily measured: time, distance, and mass. According to Einstein's relativity theory, however, most measurements are not absolute. Contrary to ordinary experience, they depend on the frame of reference of whoever is doing the measuring—that is, on the location and motions of the observer. This is especially true at speeds approaching the speed of light.

The first assumption of special relativity theory is that observers inside a uniformly moving frame of reference will perceive physical events within their frame to be unaffected by its motion. But when observers look outside their own frame, its motion will affect what they see. This is the direct result of the theory's second assumption: that the speed of light is constant for all observers in uniform motion.

Observers moving at different speeds relative to a light source thus will receive the light at different times and will be unable to agree that given events occur simultaneously. Without simultaneity we can make no comparison against a standard, such as determining the accuracy of clocks or matching both ends of a train to a length of track.

Time

An observer aboard a train with a tall clocklike device that sends pulses of light from the top to the bottom of the clock face sees the light pulse at the same rate whether the train is moving or not (*right, top*). But to an observer standing at the station, the moving train's light clock runs slow. Movement causes each light pulse to travel in the direction of motion as well as from top to bottom, resulting in a long diagonal path (*right, bottom*). Since the speed of light is fixed, the light pulse takes longer to reach the end of its path. The faster the train, the longer the path and slower the clock.

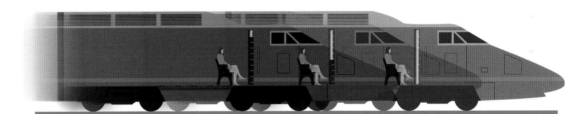

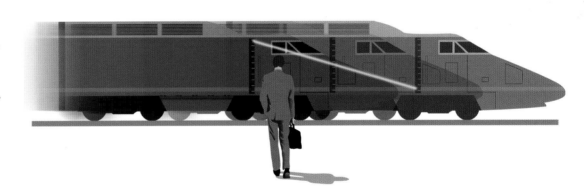

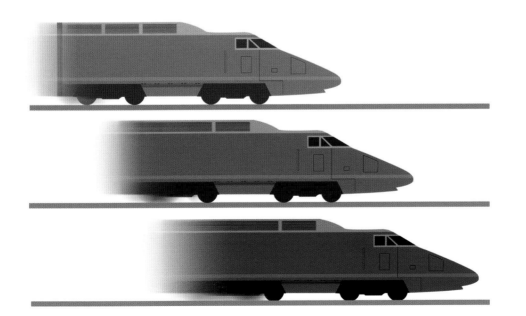

Mass

According to relativity theory, an object's mass (indicated here by degree of opacity) increases with its velocity, a phenomenon that has been verified in particle accelerators. The mass of a vehicle traveling at half the speed of light—nearly 20,000 times the normal orbiting velocity of the Space Shuttle—increases only 15 percent. (For simplicity, other relativistic effects are not shown here.) At 70 percent of the speed of light, the mass increases 40 percent. But at the speed of light an object's mass would be infinite.

Length

Two observers, one onboard a train and the other standing outside it, try to measure the train's length by comparing it to a measured length of track. A light at the forward end of the track will flash when the front of the train reaches it; the light at the back will flash when the back of the train passes. As shown below, the outside observer, standing at the mid-point of the track, sees both lights come on simultaneously and concludes that the train is exactly as long as the section of track.

The onboard passenger reaches a very different conclusion, as shown in three steps at right. The train's motion carries the passenger forward to meet the light coming from the forward signal. Because she receives that light before any light arrives from the signal behind her (*top*), the passenger concludes that the front end of the train has reached the forward mark on the track before its back end has passed the rear mark; therefore, the train is longer than the track (*middle*). Indeed, the passenger's estimate of the train's length increases until she finally perceives the flash from the rear light (*bottom*).

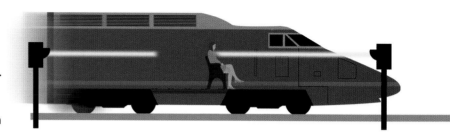

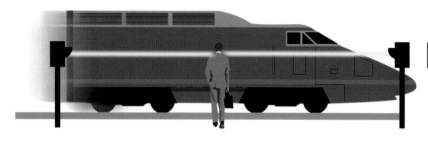

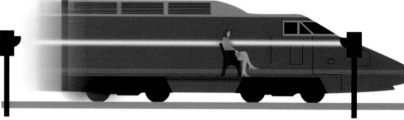

tall, uneven staircase. It can hop one step or many steps at a time, going either up or down. However, it cannot stop in midair between two steps. That's one of the odd rules that physicists discovered for electrons in energized atoms. They can orbit only at certain "levels" around the atom, and at no others.

Even stranger, we can never know exactly where the electron-kangaroo is or how much energy it contains. The more precisely we try to measure it, the more uncertain it becomes. This inability to measure the exact speeds and positions of subatomic particles at the same time is called the Heisenberg Uncertainty Principle, named for the German physicist Werner Heisenberg. It means that electrons are not bits that we can pinpoint in space, like motes of dust; rather, they are statistical blurs. Electrons and the tiny units of light energy called photons act like particles in some circumstances and like waves in others, a final oddity that defies rational explanation.

To account for all of this, scientists developed a theory called quantum mechanics. It sounds like science fiction, but one experiment after another has confirmed the ability of quantum mechanics to predict the behavior of matter. Despite its impressive string of successes, Einstein found quantum mechanics so intellectually unpleasant that he spent the last decades of his life trying to disprove it. Even today, this subatomic world could not be more alien to our senses. Nevertheless, we depend on quantum mechanics to learn the properties of objects halfway across the universe. Much of our modern technology, from photocopiers to laser pointers to supercomputers, relies on quantum mechanics. But what are the exact speeds and locations of the electrons and photons in these devices? Heisenberg knew the answer: We can never know.

The Universe GOES 'ROUND

A more tangible component of motion that appears everywhere is rotation. We could supplement our summary of the cosmos, "Everything moves," with a companion statement, "Everything spins." This applies to electrons and atoms, eddies in a stream, storms in the atmosphere, asteroids and planets, stars, and entire galaxies.

At first, this rule of thumb might not seem obvious. Anything near at hand that you could set twirling—a coin, a top, an egg—seems eager to grind to a halt fairly quickly. That's because friction resists the motion of a spinning object when it touches something else. But once you reduce friction, as Galileo did for motion in a straight line, the persistence of rotation becomes clear.

One easy way to demonstrate this is to spin in your desk chair, preferably when your boss or teacher isn't looking. A swift sideways push off any handy surface sends you whirling. But the exhilaration is short lived. Friction between the spinning metal post and the stationary base of the chair converts the rotational energy into heat, and you slow to a light-headed stop. If you coat the post with some oil, you might spin two or three times longer. The thin layer of slippery liquid lowers the friction but doesn't eliminate it entirely.

Now think of a spinning basketball. If you plunk the ball into a tub of water and spin it, the drag from countless water molecules on the bottom of the ball quickly slows it down. But if you are dexterous enough to spin the ball on your finger, you can reduce the friction to a single point of contact between you and the ball. As a result, it spins much longer and your friends walk away impressed. Jugglers also reduce friction in their routines by perching spinning objects at the ends of pointed sticks, or even on the tips of their noses.

Finally, imagine that the rotating thing touches only air. A tossed disk of pizza dough doesn't stop twirling in midair because the friction between the dough and molecules in the air is very small. In the same way, crisply struck golf balls or smartly thrown boomerangs maintain most of their spins until they fall back to Earth. If we go one step further and subtract the air itself, there's no limit to how long an object can spin. Physicists can do this in the laboratory by levitating objects with magnetic fields in vacuum chambers where almost all of the air is pumped out. In such settings, objects could rotate for years if anyone cared to watch.

But the best vacuum of all is in space. Once a body starts to spin there, it won't stop unless it collides with something or feels some source of drag, such as tidal pulls

Conserving Momentum in the Cosmos

The principle of the conservation of angular momentum explains in part why some galaxies are elliptical and others have the familiar disk shape that characterizes spiral galaxies like our own Milky Way. Just as a skater spins faster when she pulls her arms and legs in toward her body (*below*), the gas and stars that coalesce to form galaxies begin to swirl faster as they fall inward toward greater concentrations of matter.

A galaxy's final shape is determined primarily by how rapidly its gas condenses to form stars. An initially lumpy distribution of gas may lead to more rapid star formation, using up the available gas and producing an elliptical shape, with stars smoothly distributed outward from a dense core. Disk galaxies, as shown here, may have more diffuse gas to start with but then end up with leftover swirling gas that tends to converge in a disk.

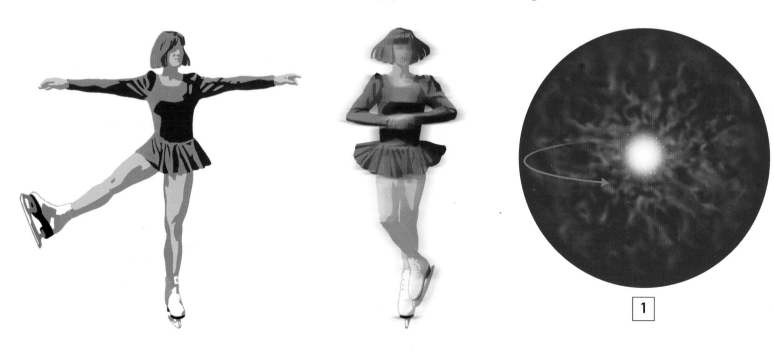

1

from nearby objects. The physical principle that governs such long-lasting motion is called the conservation of angular momentum. According to Newton's laws of motion, an object's straight-line momentum stays the same unless some outside force acts upon it. Angular momentum behaves the same way, with an added twist. The amount of momentum carried by a spinning object depends not only on its rotation speed but also on how its mass is distributed around its axis of rotation. That quantity, known as the "moment of inertia," dictates how much force it takes to spin up an object of a certain shape—and how much the object resists slowing down once it is spinning.

It's easy to see this concept in action. A familiar example is the dramatic spin used by figure skaters to end their performances. Watch as the skater spins slowly at first, then draws her arms and legs toward her body. That brings her distribution of mass closer to her axis of rotation, which lowers her moment of inertia. But because her total angular momentum must stay the same (ignoring the minor friction of the skates against the ice), her rotation speed must increase drastically to make up for it.

The initial collapse of randomly moving gas and stars forms a bright nucleus of large star clusters with a defined orbital spin (1). The upward and downward motion of the remaining diffuse gas clouds cancels out; conserving angular momentum, the clouds retain only their orbital motion around the nucleus (2).

As the central bulge grows, gas clouds gradually converge into an orbit along the equatorial plane of the galaxy (3). The resulting disk is the most stable compromise between gravity's pull and the energy of the original motion of the gas clouds around the nucleus (4).

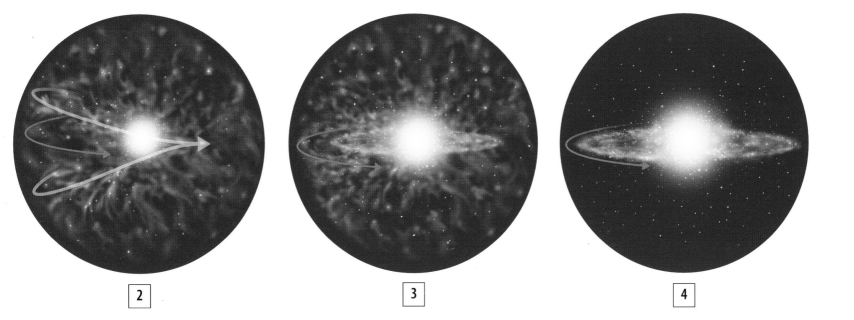

2   3   4

You can experience the same process without risking an icy fall. Simply return to that desk chair and ask a friend to spin you around, except this time, hold your arms outstretched and grasp a heavy object in each hand. Two phone books will do. After you're rotating at a healthy clip, bring the books inward toward your chest. Get ready for a dizzying whirl as your moment of inertia shrinks—and don't try to read the books for a few minutes afterward.

Such changes in moments of inertia are critical in astronomy because they explain why so many things in the universe spin rapidly. When a new star forms, clumps of gas coalesce into a body that grows progressively denser and more massive. However, the gas doesn't fall in straight lines toward the center, since a collapsing cloud in space always starts with some slight spin. Material drifts inward along curved paths as the cloud's moment of inertia decreases, making it spin more quickly. After hundreds of thousands of years, the baby star grows dense enough to ignite nuclear fusion at its core. By that time it rotates as fast as once every day, a breakneck pace for a ball of gas a million miles wide.

Planets spin for the same reason. They form when smaller clumps of matter congregate within the dusty disk that remains around a new star. The disk revolves in the same direction as the star, resulting in planets that spin in the same direction as their sun. However, there are two exceptions in our solar system. Uranus has an axis of rotation tilted so severely that it rolls on its side like a gaseous bowling ball. Venus spins slowly in the opposite direction, a property called "retrograde" spin. Fierce collisions or mergers with other large bodies early in the solar system's history probably created these curiosities.

Even the planets with well-behaved spins aren't perfect rotators. Earth's axis tilts 23.5 degrees from the plane of its motion around the Sun. This leads to variations in the angle of sunlight striking the ground, creating our cycle of seasons as the length of daylight changes. The rotating Earth also exhibits another more subtle behavior. The axis of a tilted rotating object wobbles in a circle if another force acts upon it, a process called "precession." For a child's spinning top, that force is Earth's gravity. The top wobbles slowly at first, then more quickly as friction slows it down and brings it tumbling to the floor. For Earth the force is the combined gravity of the Moon and the Sun. The planet's axis takes 26,000 years to precess through one circle. Today,

Such changes in moments of inertia are critical in astronomy because they explain why so many things in the universe spin rapidly.

Variations on Rotation and Revolution

All the planets in the solar system obey the physical laws of rotation and revolution, but their individual cycles vary widely. Mercury, for example, has been greatly affected by its nearness to our home star. The Sun's immense gravity has exerted a strong tidal pull, slowing Mercury's rotation until the planet takes almost 59 Earth days to complete one full spin on its axis. This snail-like rotation, combined with the planet's highly elliptical orbit and the fact that its counterclockwise revolution around the Sun takes the equivalent of only 88 Earth days, has odd consequences, as shown here. Mercury is also unique in that its axis of rotation is not tilted with respect to the plane of its orbit. Earth, for example, is tilted 23.5 degrees to its orbital plane, and Uranus, with an axial inclination of 82.1 degrees, wallows around the Sun on its side (page 30).

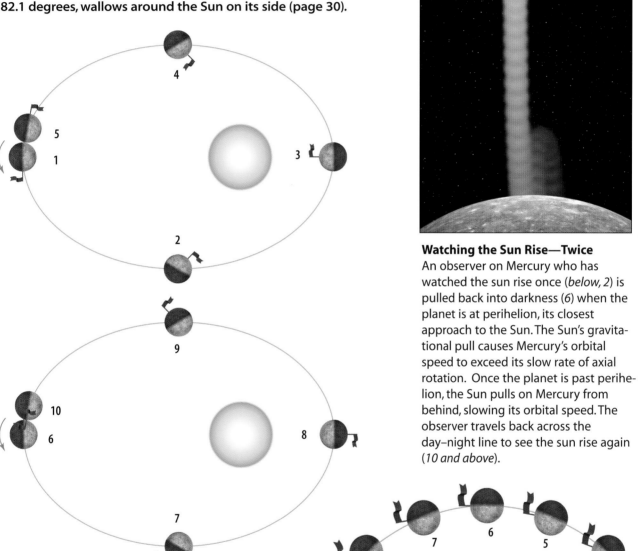

Watching the Sun Rise—Twice
An observer on Mercury who has watched the sun rise once (*below, 2*) is pulled back into darkness (*6*) when the planet is at perihelion, its closest approach to the Sun. The Sun's gravitational pull causes Mercury's orbital speed to exceed its slow rate of axial rotation. Once the planet is past perihelion, the Sun pulls on Mercury from behind, slowing its orbital speed. The observer travels back across the day–night line to see the sun rise again (*10 and above*).

A Long (Two-Year) Day

Because of its 59-Earth-day rotation and 88-day orbital period, Mercury's days last two years (*above*). If an observer (*flag*) sees the Sun come up when Mercury is at aphelion, farthest from the Sun (*1*), noon occurs at perihelion (*3*). Sunset (*5*), at the next aphelion, ends one year and begins the next (*6*). Midnight occurs at the next perihelion (*8*). Dawn (*10*) begins year three.

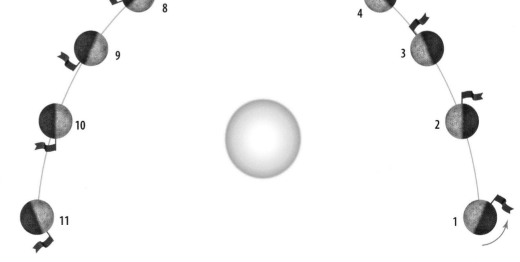

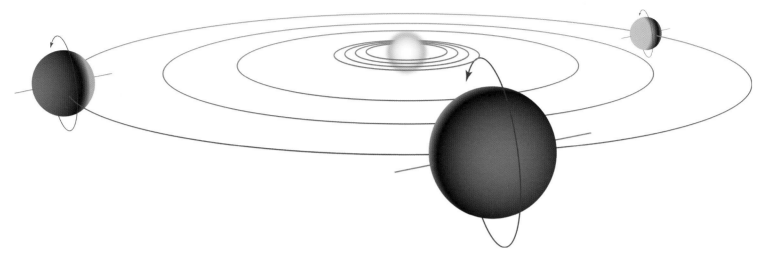

A Severely Tilting Planet

Most of the other planets in the solar system spin more or less like tops as they travel around the Sun, but Uranus tilts severely, circling with its axis of rotation nearly parallel to its orbital plane. One result of this orientation is that its polar regions alternately point directly at or away from the Sun. In the course of the planet's 84-Earth-year orbit, the poles endure 42-year days and nights. Scientists think that an Earth-sized object might have smashed into Uranus early on, knocking it on its side.

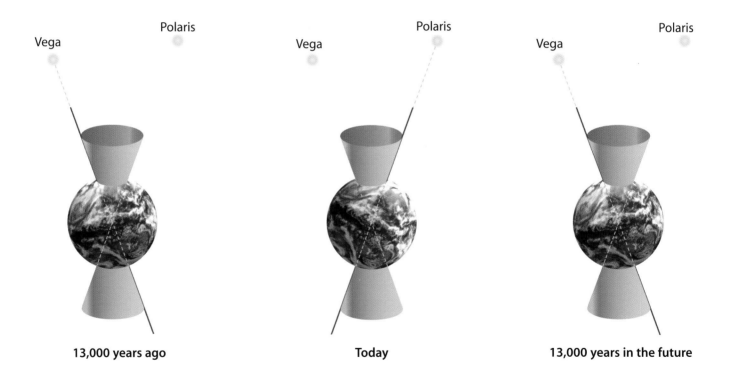

Earth's Precession

The gravitational pull of the Sun and Moon on Earth's equatorial bulge causes Earth's rotational axis to gyrate around the vertical, tracing a cone in space over a period of 26,000 years. The pole star to which the axis points today is Polaris. Thirteen thousand years ago Earth's axis pointed to Vega, in the constellation Lyra, as it will again 13,000 years from now. Precession was the third-discovered motion of the Earth, after the far more easily observed daily rotation and annual revolution.

Earth's axis points in the northern sky at Polaris, a moderately bright star at the end of the Little Dipper's "handle." That will change as precession swivels Earth's axis on a slow arc through the sky. In 13,000 years, Earthlings will have a dazzling beacon as their north star, the bright star Vega.

Rapid spin imparts another desirable property to an object. In the absence of other forces, it keeps the axis of rotation pointing in the same direction. To use a recreational analogy, try to hurl a Frisbee without spinning it. It will flutter embarrassingly to the ground. But a quick snap of the wrist sends the disk on a steady rotating flight through the air. Spinning bicycle wheels also are much more stable than stationary ones, as any cyclist can attest. Spacecraft designers take advantage of this property in two ways. Some satellites and space probes spin several times each minute, good enough to keep their communications antennas aimed at Earth and to prevent them from tumbling through space. Orbiting telescopes require a more accurate aim. Those spacecraft carry at least three small gyroscopes, each aligned perpendicular to the other two. Guidance systems use the combined motions of the gyroscopes to hold the satellites still, allowing mission controllers to point them with great accuracy.

Like spinning basketballs and satellites, rocky planets are rigid rotators. To maintain their shapes, all parts of the objects make complete turns in the same amount of time. This means that a point on the surface has zero speed at the axis of rotation. At the object's equator (the belt most distant from the rotation axis) the surface speed reaches a peak. On Earth that's 1,040 miles per hour, modest by cosmic standards. Most launch pads are close to the planet's equator, rather than the poles, to give rockets an extra boost of speed on their climbs into orbit.

Earth and other rigid rotators are strong enough to withstand the internal stresses that otherwise would shear the objects apart as they spin. However, the universe is full of rotating bodies that aren't rigid at all. A star is one example—it's a bloated ball of gas with little stiffness. A stirred mug of coffee is another. It's easy to notice that the coffee near the center of the mug rotates at a rate altogether different from the coffee near the rim. If you need a visual aid, dribble some cream from the mug's center to the edge. If the coffee rotated as one solid unit, like Earth, the cream would remain intact in a fuzzy line. Instead it stretches into strands and loops that spin at varying rates. This shearing action is called "differential rotation."

Spinning Storms and Worlds

The Coriolis effect is an apparent deflection in the paths of moving objects caused by Earth's faster speed near the equator than near the poles. In the solar system, the greatest expression of this effect occurs in planetary atmospheres. On Earth, for example, the air around us moves freely above the ground. Low-pressure centers draw air in; high-pressure centers push air away. As illustrated here, Earth's differential rotation bends the paths of moving air just enough to create a consistent circulating pattern.

Unlike spinning bicycle wheels and other kinds of rotation we see all around us, Earth's rotational rate is very small—only one rotation a day. Water swirling down a drain, by contrast, may take only a few seconds to make one rotation, a very fast rate. Thus, contrary to popular belief, the Coriolis effect doesn't influence the direction of draining water one way or the other.

The Coriolis Effect

Because Earth rotates faster near the equator than near the poles (indicated by the varying lengths of the yellow arrows), moving air drawn to low-pressure areas at midlatitudes travels east either faster or slower than the low itself. As the low draws air in (white arrows), the difference in speeds causes the air to curve—counterclockwise in the Northern Hemisphere and clockwise in the Southern Hemisphere (purple arrows).

Not Coriolis

The direction of rotation of a draining tub is determined by the way it was filled or by ripples introduced while washing. The magnitude of these rotations may be small, but at the scale of your bathtub they easily overpower any effect of Earth's rotation.

On Earth

As shown in this photograph of the 1997 Pacific storm named Hurricane Linda, air moving from all directions toward a low-pressure area rotates counterclockwise in the Northern Hemisphere. By contrast, air being pushed away from high-pressure areas circulates clockwise north of the equator and counterclockwise south of the equator.

In the Solar System

Jupiter's Great Red Spot, a huge storm bigger than Earth itself, exemplifies the Coriolis effect at work on other planets. Observed for more than 300 years, the Great Red Spot rotates counter-clockwise in Jupiter's Southern Hemisphere—a dead giveaway that it is a high-pressure system.

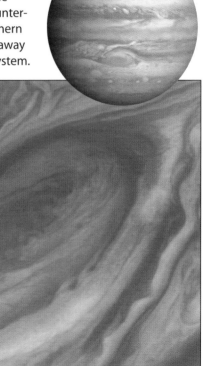

Earth's atmosphere rotates in this way because it is not solidly connected to the surface of the planet. This sets up patterns of high-altitude winds that drive weather systems in certain directions. Over the United States, those winds usually blow from west to east, allowing us to track storms as they move across the country. Other parts of the globe see winds in the opposite direction or, near the equator, meandering breezes known as the doldrums.

Another consequence of combined motion and rotation affects our lives dramatically on Earth and creates striking patterns on other worlds. We call it the Coriolis effect: an apparent deflection in the paths of moving objects caused by Earth's faster speed near the equator than near the poles (pages 32–33). Low-pressure systems in the Northern Hemisphere—including hurricanes and tornadoes—tend to rotate counterclockwise as a result of these deflections. In the Southern Hemisphere the patterns are reversed.

According to urban legend, the Coriolis effect compels water to drain from tubs and toilets either clockwise or counterclockwise in each hemisphere. That's a fun myth, but it's not true. The direction that water flows out of a basin is determined almost entirely by the basin's shape and random currents in the water. You'd see evidence of the Coriolis effect only if the basin was many miles across and totally still when you pulled the plug.

The Coriolis effect has the same result as a sideways push on a moving object. At its strongest on Earth's surface, the force is just one three-hundredth as strong as the planet's gravitational pull. A football kicked 50 yards either due north or south would move sideways about half an inch thanks to this effect—probably not enough to change the outcome of a game. But when something travels many miles, the effect becomes substantial. In an undistinguished moment in the history of warfare, British naval forces forgot to adjust for the reversed direction of the Coriolis effect in the Southern Hemisphere when they clashed with German warships near South America during World War I. As a result, about a thousand missiles missed their targets and fell into the ocean.

GRAVITY'S Hold on the Cosmos

We might like to attribute the lovely spirals of rotating galaxies to the Coriolis effect as well. However, the physical processes are very different. Spiral arms probably arise from shock fronts called "density waves" that ripple through the gas in a galaxy. When stars and gas encounter these waves, they move more slowly than they do in the rest of the spinning galaxy. This piles them up in galactic traffic jams that generate bursts of newly formed stars within the gas clouds. We see the denser concentrations of stars as spiral arms. Cars moving on a congested highway clump together in a similar way. Dense knots of traffic form as the speeds of vehicles vary. A single slow-moving truck (*right*) can have a ripple effect, forcing vehicles to slow down and speed up in frustrating bursts for miles behind. These knots gradually creep along the road. The knots themselves may persist for a long time, even though individual cars get stuck in them for only a few minutes.

The force that dictates the shapes of those spiral arms is gravity—the same force that holds you in your chair as you read this page. We owe our initial understanding of gravity to Isaac Newton. His three laws of motion alone would have guaranteed his

Of Spiral Arms and Traffic Jams

Spiral galaxies may resemble terrestrial hurricanes, but the forces that create their distinctive spiral arms are quite different. Indeed, the differential rotation that produces the Coriolis effect (and hurricanes) should eventually disperse any spiral structure that might form in a galaxy. One mechanism to explain how spiral arms persist is a spiral-shaped density wave that moves through the galactic disk more slowly than the orbiting stars, dust, and gas. As gas and dust get caught in the galactic traffic jam, the congestion triggers a burst of star formation that lights up the spiral pattern for a time (*below*). Then, like the knot of cars that repeatedly bunch up around a slow-moving truck (*right*), the density wave moves on.

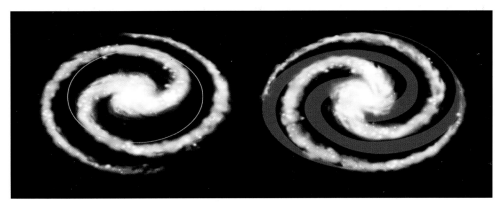

Density Waves

When orbiting gas and dust (*yellow circle, above left*) intersect a spiral density wave, the resulting burst of new star formation lights up the spiral. Because of differential rotation, older stars move ahead of the density wave, creating the trailing arms. As older stars die, the original pattern smears out. Millions of years later, the density wave has rotated, the new spiral arms are lit by different stars, and the old arms fade away (*above right*).

lasting fame as a scientist, but he also is justly renowned for deriving a powerful equation known as the law of universal gravitation.

The genesis of that insight makes for perhaps the most famous tall tale in the history of science. As Newton sat under an apple tree and watched the Moon, the story goes, an apple fell from its branch and struck him on the head. At that point, Newton realized that the force drawing the apple to the ground and the force keeping the Moon in its orbit around Earth were one and the same. (Most historians believe that Newton did indeed see an apple fall to the ground, but it almost certainly did not hit his head.) No matter how it actually happened, Newton was the first scientist to recognize that the laws of nature link everything above, beneath, and around us into one universe.

Newton explained that gravity pulls toward a center, not just down to the surface. In other words, the apple fell toward Earth's center, and Newton's head happened to get in the way. His equation shows that gravity's strength decreases with the square of the distance between two objects. If the distance between the center of the Sun and a comet is tripled, the gravitational force between them is one-ninth as strong. Further, the force depends directly on the mass of the object doing the pulling. Mars would attract the apple with less force than Earth because of that planet's lower mass, whereas Jupiter's pull would be much greater. If you've ever stepped on one of those scales at a planetarium that reveals how much you would weigh on another planet, you've participated in a calculation of the law of universal gravitation.

That equation and Newton's second law of motion predict that gravity will accelerate two things of different masses at exactly the same rate. Galileo demonstrated this nearly a century before Newton by rolling and sliding smooth objects down inclined tracks. In a tale that likely has a kernel of truth, he also dropped cannonballs of unequal masses from the Leaning Tower of Pisa to prove that they hit the ground at the same time. *Apollo 15* astronaut David Scott dropped a hammer and a falcon feather onto the Moon's surface in 1971 to show the same thing. (Scott's trick won't work in your living room because air resistance on Earth makes a feather flutter rather than fall.) Indeed, the practical significance of gravity's equal acceleration of objects is critical in the space program. For instance, NASA engineers don't need to

Hot blue-white stars light up the graceful spiral arms of Galaxy NGC 2997.

Of the four basic forces of nature—gravity, electromagnetism, and the "strong" and "weak" forces that operate within the nucleus of an atom—gravity is by far the weakest.

worry about the mass of a space probe when they use a planet's gravity to adjust its flight path. A heavy spacecraft and a light one respond in exactly the same way.

It's hardly necessary to review why gravity is important in our lives. However, most people don't appreciate something surprising about it: Of the four basic forces of nature—gravity, electromagnetism, and the "strong" and "weak" forces that operate within the nucleus of an atom—gravity is by far the weakest. To see this in one way, just rub a balloon in your hair or on a wool sweater a few times and stick it on your sleeve. Despite the gravitational pull of the entire Earth, the balloon stays attached to your arm because of a tiny bit of static electricity. On a far smaller scale, the strong force that binds the nucleus of an atom together is more powerful than gravity by a factor of more than a trillion trillion trillion. And yet gravity dominates every facet of the universe, from the Big Bang to the creation of galaxies, stars, and our planet.

This seems strange, but the reasons are easy to comprehend. Two of the four basic forces work only over the tiny distances within atoms. Electrical attraction is powerful over larger distances, but matter contains positively and negatively charged particles (protons and electrons, respectively) that cancel each other out. The only force that has no negative counterpart yet still operates across vast reaches of space is gravity. In matters universal, gravity always wins.

Another way to look at the dominance of gravity is to consider what happens when Earth's gravity accelerates falling objects. When you jump from a bridge with a bungee cord around your feet, your speed reaches about 20 miles per hour after one second and about 40 miles per hour after two seconds. A powerful motorcycle also can accelerate that fast on a straight, smooth road. But the motorcycle's engine does not rival Earth. It can accelerate only one vehicle, whereas Earth could accelerate millions, even billions, of bungee jumpers at the same time, if that many people took leave of their senses and jumped at once. That is the power of gravity: On small scales it cannot compare with forces generated by other means, but on planetary, interstellar, and galactic scales, it has no equal.

We are still learning about gravity's effects on objects in the cosmos. For example, stars in the outer parts of spiral galaxies revolve surprisingly quickly around the galactic centers. We believe that an unseen extra source of gravity—some mysterious "dark matter"—exerts an additional tug on the stars as they orbit. This tug pulls the stars along faster than they would travel if the gravity came only from the glowing stars themselves. Dark matter will play a key role in determining the fate of the universe billions of years from now.

The oddest thing about gravity is that it acts across the gulf of space with no apparent connection between objects. Even an apple lacks an obvious reason to fall down. There's no contact between it and the Earth, and no easily identified "field" like the electromagnetic fields around bar magnets and electrical transformers. When Albert Einstein considered this situation and other riddles about gravity, he devised his general theory of relativity—a profound idea that, once again, extended Newton's work.

One part of the theory, published in 1916, holds that there is no difference between the acceleration due to gravity and the force felt within an accelerating frame of reference. Imagine standing in an enclosed box and feeling a downward force that produces your normal weight. The box could be resting on Earth's surface, and the downward force you feel would be the planet's gravitational pull. However, the box also could be zipping on a straight line through space with a constant acceleration of 32 feet per second every second. That would produce a force toward the floor of the box identical to Earth's pull. The box could even be whirling at the end of a long cable in space or inside a giant rotating wheel, like the spacecraft in *2001: A Space Odyssey*. Inside your box you could not distinguish among these situations. The next time you feel your weight "increase" as an elevator starts upward, remember that you'd feel the same sensation if Earth's mass suddenly increased to create a stronger gravitational pull—or if your spaceship's engines just turned on.

The most mind-bending impact of general relativity came from the way in which Einstein changed our conception of space and time. His theory held that throughout the cosmos, space and time are woven together in four dimensions, like the threads in a piece of cloth. Gravity, Einstein proposed, is the wrinkling of the fabric of this "space-time" by objects within the universe. That statement sounds unfathomable. However, it helps to think of a three-dimensional analog to Einstein's idea. Touch your fingertip to the surface of a bowl of Jell-O. The surface bends inward with the weight of your finger, the same effect that a mass creates in the weave of space-time.

The simplified model below illustrates Einstein's statement that massive objects warp the fabric of four-dimensional space-time. If space-time is viewed as a sheet, then the gravity of objects such as galaxies, stars, and planets wrinkle the sheet to varying degrees. Other objects must follow the resulting bends and curves—including light.

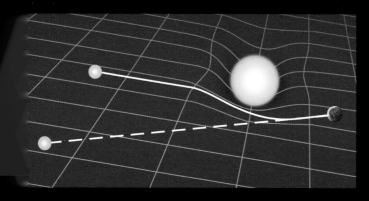

As light from a distant star passes by the Sun on the way to Earth, it follows the curved path (*solid line*) resulting from the pull of the Sun's mass on the space-time sheet. Einstein said that the star's apparent position (*dashed line*) would differ from its real position by a predictable amount.

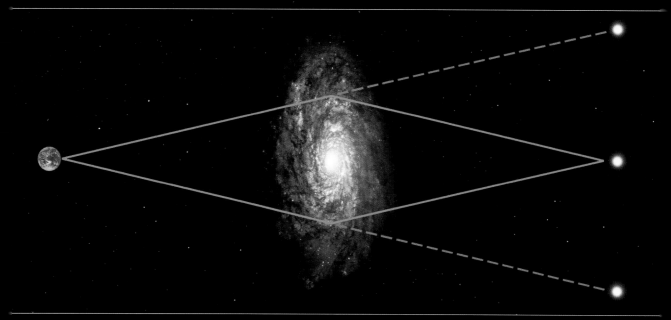

Gravitational Lensing

The gravity of a massive object bends light from more distant galaxies if the objects are closely aligned in space. As shown by the dashed lines in the simple model above, this projects the light into multiple images. Such lenses act like giant magnifying glasses, creating exotic patterns of arcs and rings. The orbiting Hubble Space Telescope has revealed many of these cosmic mirages. Light from a distant galaxy is smeared into several ghostly arcs (*near right*) by a massive galactic cluster. When the alignment of the lensing galaxy and the distant object is nearly perfect (*far right*), the result is a rare circle of light called an "Einstein Ring."

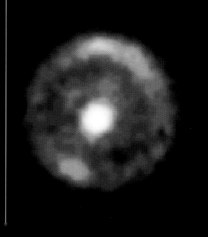

According to Einstein, living in the universe is like living on a huge piece of soft elastic rubber. Space-time is a medium that has shape and form. Objects within this medium can flex and twist it. Every object in the universe pulls on the space around it, drawing the fabric of space-time toward its center. The more massive the object, the more it pulls. The amount of pull exerted by an object on the universal fabric is its gravitational force. So the apple falls to Earth because Earth has warped space-time in such a way that the apple must move toward Earth's center. More massive planets create a deeper warp, imparting a faster acceleration to objects that wander past. The physicist John Wheeler captured these odd notions perfectly: "Matter tells space how to curve, and curved space tells matter how to move."

Gravity and LIGHT

Just as remarkably, gravity works not only on objects but also on light. After all, like apples and cannonballs, light travels through space-time, too. A massive body, such as a star, bends a passing beam of light much as a subtle curve in a putting green bends a golf ball toward the hole. Einstein calculated exactly how much our Sun would deflect starlight in this way. He predicted that the effect should be noticeable during a solar eclipse. In 1919 the British astrophysicist Sir Arthur Eddington set up expeditions to Africa and South America to look for changes in the positions of stars near the Sun during a total eclipse. The displacements were extremely small, an angle less than the thickness of a dime as viewed across a football field. Even so, they matched Einstein's prediction. At that moment, Einstein became an international celebrity. Such a displacement of light is called "gravitational lensing." It occurs throughout the universe. Massive objects act like handheld glass lenses, magnifying and distorting light waves from the stars and galaxies behind them.

Other effects of gravity are considerably more down to Earth. One that everyone recognizes is the daily swelling and retreat of Earth's oceans, known as tides. Tides occur because the Moon's gravitational pull is strongest on the side of the planet nearest the Moon, less strong at Earth's core, and weakest on the far side of the planet. The differences in these forces are enough to distort Earth and its oceans into a slightly elongated shape, pointed in the direction of the Moon. It's hard to notice the bulging of the solid Earth, but oceanic flows produce high and low tides twice a day as Earth

Tidal Forces

All points on worlds in orbit around one another feel the gravitational pull of the other body. However, the sides facing each other feel the most pull (*longest arrows, right*), whereas the far sides feel the least. These differential tidal forces distort the round shape of planets and moons and are responsible for the tides on Earth. A high tide occurs on the side of Earth facing the Moon because the Moon pulls more strongly on water there than it pulls on the center of Earth itself. And because the Moon tugs on water on the other side of Earth even less than it does on the planet's center, a high tide also occurs on the other side (*right*). The Moon's surface bulges three times higher than does Earth's, however, and Earth's pull has slowed the smaller body's rotation so that the Moon always shows us the same face (*opposite*).

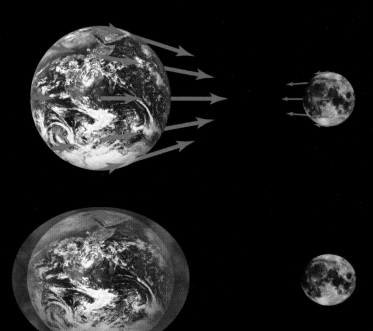

Gravitational Torture

The same forces that cause tides on Earth squeeze and stretch Io, the innermost of the four Galilean moons of Jupiter. Gravitational forces generate enormous frictional heat in Io's interior, much as a tennis ball warms up if it is repeatedly squeezed. The moon's multicolored surface (*left*) is the product of this inner roiling: Volcanoes dwarfing any on Earth remake Io's surface with relentless new lava flows and deposits of sulfur dioxide.

Tug-of-War

Io suffers extreme tidal flexing because of the competing pulls of its giant planet and its three neighboring moons (*left*). In one 41-hour orbit, parts of Io's surface can rise and fall more than 300 feet, the equivalent of a 30-story building.

spins beneath the Moon. The Sun also raises tides on Earth but only about half as effectively because it is much farther away. The large distance leads to a relatively small difference in the Sun's gravitational force between one side of Earth and the other.

Billions of years of tidal pulls between Earth and the Moon have altered their orbits and rotations. Tidal friction within the Moon, part of which was molten early in its history, slowed its spin significantly. Today, it rotates just once each time it circles Earth. We call that phenomenon "tidal lock." It's the reason that we see the same face of the Moon each night. The Moon slows Earth's spin as well but much more gradually because of Earth's larger angular momentum. Still, a day on Earth was much shorter in the past. For instance, growth rings in fossilized corals indicate that 400 million years ago each day was just 22 hours long. The length of our days now increases at the rate of 16 seconds per million years, still not enough extra time for us to get everything done.

The same tidal forces happen with more dramatic results around Jupiter. The moon Io, the closest of Jupiter's Galilean satellites, is the most tidally tortured object in the solar system. The combined gravitational pulls of Jupiter's other major moons—primarily Europa, the next nearest—tug Io to and fro in its otherwise circular orbit. Massive Jupiter thus inflicts powerful and ever-changing tidal stresses on Io. This churns the moon's interior and melts it, just as you could liquefy a bag of ice

Tidal Lock

Earth's much greater gravitational pull has slowed the Moon's axial rotation so that it now spins once in the same amount of time it takes to complete one orbit of Earth: 27.32 Earth days. Thus we always see the same side of the Moon (*left*), with its dark maria, or plains of ancient lava flows. The less familiar far side (*right*), photographed by astronauts and space probes, shows the cratered history of meteoritic bombardment.

After hundreds of millions of years, the two galaxies may merge into a disturbed and disrupted blob. This scenario may be in our future.

cubes by shaking it violently. Io is scarred by fresh eruptions of molten sulfur and other minerals, which completely repave the moon's surface every few dozen years.

The next moon outward, Europa, also is heated by Jupiter's tides, but less vigorously. It may be warmed enough to maintain a deep ocean of liquid water beneath its icy cracked crust. This is tantalizing, for many biologists believe that life on Earth began in the oceans. If the right conditions exist, Europa's dark seas could even now be nurturing extraterrestrial life—without the Sun as a source of energy.

Some of nature's most spectacular displays of cosmic mayhem arise from tidal interactions on a much larger scale. Entire galaxies can collide, drawn together by their mutual gravitational pulls. The gravity of each galaxy tugs much more strongly on stars in the nearer side of the other galaxy. These tidal forces stretch the original structures into sweeping arcs and rings of stars. After hundreds of millions of years, the two galaxies may merge into a disturbed and disrupted blob. This scenario may be in our future. Our Milky Way and its neighbor, the Andromeda galaxy, are moving toward each other and could meet in about 5 to 7 billion years. By that time it won't matter for Earth because our Sun will have run out of its nuclear fuel.

Individual stars almost never smash together when galaxies collide. They are simply too far apart and too small compared with the vast distances between them. If there were four snails running loose in the continental United States for a billion years, two of them would be more likely to bump into each other than would two stars during a galactic collision. However, the oversized clouds of gas and dust in galaxies certainly do interact. The gravitational turmoil stirred by this process sets off new bursts of star formation, lighting up galaxies like a holiday display. We often see the brilliant blue flares of newborn stars along graceful "tidal tails" flung into space during galactic collisions.

On the largest scale of all, the combined gravitational pulls of every object in the universe act to slow down the expansion of space-time that Edwin Hubble discovered in 1929. For years astronomers debated whether the universe contained enough mass to slow its own expansion to a halt. If so, gravity would pull everything together in the ultimate "big squeeze." But it now appears that won't happen. Our universe probably is destined to expand forever, providing more than enough time for all the stars to burn out.

Galactic Tides

For the effects of gravitational attraction at the cosmic scale, we have only to look at the fanciful shapes of celestial objects such as the aptly named Antennae (*right*) and Cartwheel (page 46). Galaxies rarely exist in splendid isolation but instead travel through the universe interacting to varying degrees with other star systems. These interactions range from stressful near misses to head-on collisions, often with spectacular results.

   Shown below are computer simulations that suggest how some so-called peculiar galaxies came to be. By calculating the gravitational interactions among each galaxy's stars, dust, and gas, scientists can reproduce such effects as the bridges of stars between galaxies and the streaming stellar tails that give the Antennae their name.

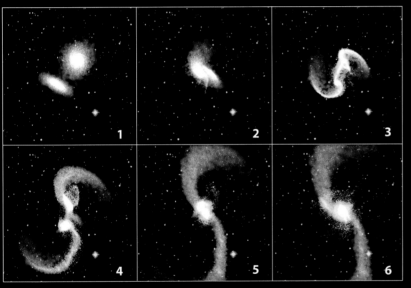

Creating Tidal Tails

This model of two disk galaxies colliding traces the gravitational effects on stars (*yellow/white*) and interstellar gas (*blue*) over about 1 billion years. The collision distorts the galaxies into an amorphous smudge (*2*). As the galaxies move apart, tidal tails curve away (*3*). Turning around on their orbit, the galaxies fall back toward each other (*4*), until they recollide (*5*) and merge into a single object that resembles the pair of galaxies known as the Antennae (*above*).

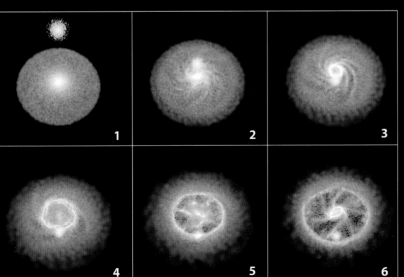

A Galactic Bull's-Eye

In this model, spanning about 350 million years, a small galaxy shoots through a larger disk galaxy along the disk's rotational axis (*1, 2*). Like the ripples caused by a rock tossed into a lake, a wave of energy expands outward, pushing gas and dust ahead of it into a ring (*4*), leaving a burst of star formation in its wake. As clumps of gas fall back toward the disk's center, they create radial spokes and an inner ring (*5, 6*), much like those seen in the Cartwheel (page 46).

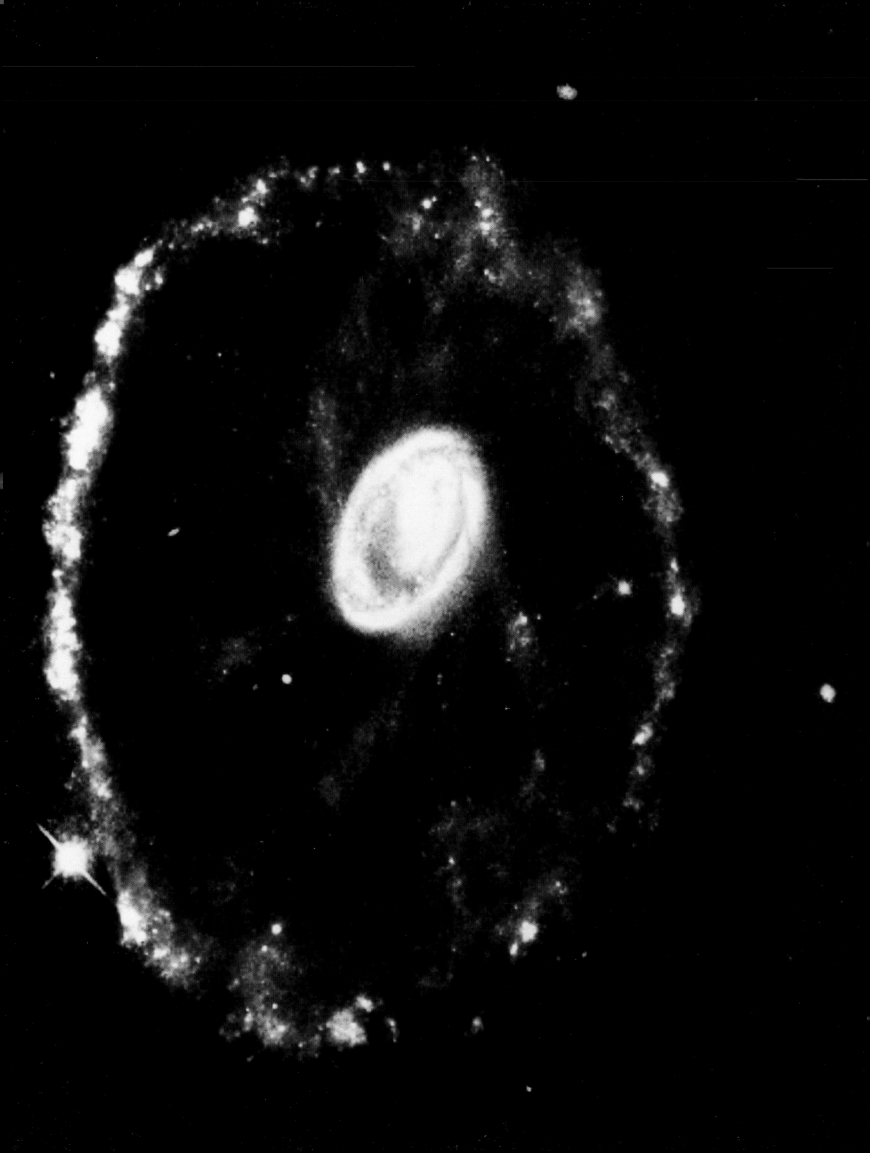

A Gallery of Peculiar Objects

The galaxies shown here are only a handful of the many varieties to be found in the universe, all products of cosmic collisions and close encounters. In the polar-ring galaxy above (*center*), the ring rotating at right angles to the plane of the disk suggests that it may be the remains of a different stellar system. Faint spokes in the Cartwheel (*opposite*) may indicate the post-collision reemergence of spiral arms.

Newton's Thought Experiment

Sir Isaac Newton began his cogitation on gravity and orbits with the fact that the pull of gravity causes a thrown object to fall in a curve. In the drawing at right, he illustrated that an object hurled from a hypothetical mountain V would fall to D. If the object could be thrown harder and harder, it would fall to E, F, or G. Finally, if it could somehow be thrown hard enough from the mountaintop, or from a point in space, it would "fall" around the world (outer circles). In following his thoughts to their logical conclusion, Newton developed the theory of orbital flight.

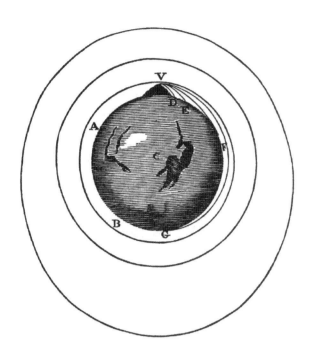

The Eternal Free Fall of ORBITS

The future of the cosmos probably was not on Isaac Newton's mind as he pondered the relation between a falling object on Earth and the Moon in orbit. He did, however, conduct a thought experiment. It went something like this: If he dropped a ball, it would descend toward Earth in free fall until it hit the ground. Launching that ball parallel to the ground from a cannon would propel it some distance, but Earth's gravity would still pull it downward in free fall. Hauling the cannon to a mountaintop would let him shoot the ball still farther. But what if he climbed the highest mountain and fired the ball fast enough—5 miles per second—so that it never touched the ground as Earth's surface curved beneath it? The ball would still fall toward Earth's center, but its fast sideways motion would keep it in a low "orbit." With this intuitive leap, Newton realized that all orbits in the solar system are never-ending free falls. Their motions are determined by the same laws that govern the flights of baseballs, the trajectories of rocks belched from a volcano, and the paths of other freely moving projectiles on Earth.

Newton acknowledged that many predecessors had set the stage for his insights. One was the German astronomer Johannes Kepler, who worked out the correct mathematical details of planetary orbits for the first time in 1609. His painstaking calculations and the observations of his mentor, the Danish astronomer Tycho Brahe, enabled Kepler to derive three key principles. First, planets move around the Sun not in

Unlike the tame orbits of the planets, most of which circle the Sun in paths that hug the ecliptic (Earth's orbital plane), comets such as Comet West (*left*) sweep through the solar system in orbits of all inclinations and orientations.

circles, as Copernicus had thought, but in oval paths called ellipses. Second, planets move faster when closer to the Sun and slower when farther away, in such a fashion that their motions sweep out equal areas of their ellipses in equal times. Third, the orbital period of a planet—its year—depends predictably upon its distance from the Sun. We can apply those laws anywhere in the universe where planetary systems revolve around other stars. We can also use them to understand other basic systems, such as two stars that orbit each other closely. But the laws are too simplistic to hold in star clusters, galaxies, and groups of galaxies, where the complexities of gravity's dances require stronger analytical tools.

Newton showed that his laws of motion and gravitation lead to other families of trajectories in the solar system besides ellipses: parabolas, hyperbolas, and, of course, circles. (Gravitational nudges from other bodies in the solar system prevent any object from orbiting the Sun in a perfect circle.) These are "conic sections"—

The Mechanics of Orbits

A major scientific breakthrough occurred in 1609 when the German mathematician Johannes Kepler published his theory that the planets orbit the Sun in elliptical, rather than circular, paths. Kepler also showed that as orbiting bodies make their closest approach to the Sun, they speed up, and then slow down as they move away. This effect explains why a planet travels from point A to point B in the same time that it takes to cover the much shorter span between C and D (*right*). Because the areas shaded blue are equal, the concept is described as "sweeping out equal areas in equal times." Building on Kepler's work, Newton showed that other types of trajectories are possible (*below*), as borne out by the orbits of the moons of Jupiter (*opposite*) and the eccentric, or extremely elliptical, paths of comets.

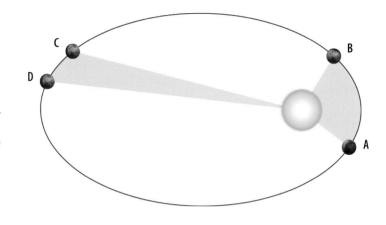

Conics

Newton's laws of motion and gravitation showed that under the influence of gravity all planets, moons, comets, and any kind of projectile will follow paths that can be described as conic sections—cuts made by the intersection of a flat plane and a cone.

 Circle

 Ellipse

 Parabola

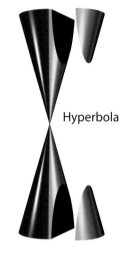 Hyperbola

shapes you can produce by slicing a cone with a sharp plane at various angles. There are as many different shapes of orbits in our solar system as there are objects. Some comets swoop in from the depths of space on near-hyperbolic paths, never to be seen again. Millions of icy particles orbit within Saturn's rings, each tracking its own near-circular course while bumping gently into its neighbors on occasion. The orbits of Mercury and Pluto are noticeably elliptical, whereas that of Venus is almost circular.

These orbits are displays of general relativity in action. The Sun creates a huge bowl in the fabric of space-time. Earth and the other planets travel along the banks of this bowl, much as marbles revolve around the sloped outer rim of a roulette wheel. The planets have just the right amount of sideways motion to keep them from spiraling into the center of the bowl or slipping out of it entirely. Earth, for instance, travels at an average speed of about 66,000 miles per hour. Its distance from the Sun varies in a stable manner between 91 million and 95 million miles. But it's not hard to imagine that smaller bodies, such as distant comets or the thousands of asteroids

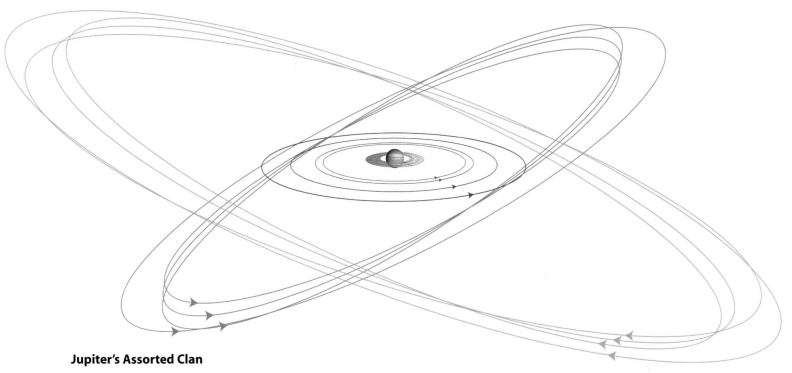

Jupiter's Assorted Clan

The moons that orbit Jupiter offer a sample of orbital variety. Eight regular moons, probably born of the disk-shaped nebula that circled the protoplanet, travel in rounded orbits in the planet's equatorial plane (beige and blue). Two groups of irregular moons (red and orange) orbit at huge distances from the planet in eccentric orbits highly inclined to Jupiter's equator, a sign that they are captured asteroids. The outermost irregulars have retrograde orbits, traveling in the direction opposite to Jupiter's own rotation.

between Mars and Jupiter, can travel more erratically. Indeed, we have learned that wayward travelers zip through our bowl in space-time with alarming frequency.

Gravitational interactions among the many bodies in the solar system, large and small, lead to long-term unpredictability in the orbits of objects. The physical principle behind these changes is called chaos. When a system is chaotic, we can only predict its motion a short time into the future. After that, even the tiniest initial changes in an object's velocity or position result in drastically different outcomes. Weather patterns in Earth's atmosphere are chaotic, which explains why forecasts aren't useful beyond a week or so. In the solar system, the combined tugs of the planets and other objects are extremely difficult to calculate. Given enough time, they will perturb an asteroid or a moon into an entirely new orbit.

It comes as no surprise that Jupiter is our solar system's gravitational bully. Close encounters with this giant planet can eject objects from the Sun's grasp or send them into our neighborhood. We know of dozens of asteroids' orbits that cross Earth's or

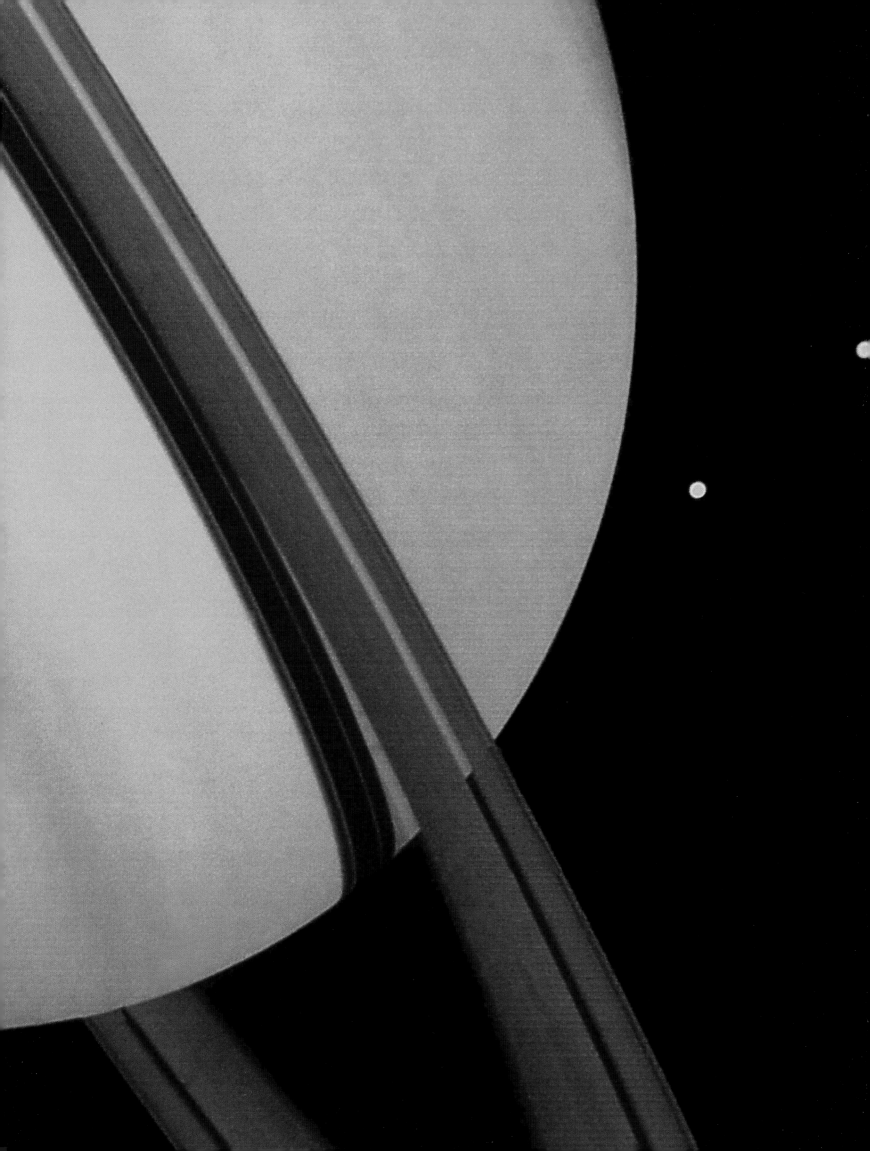

Saturn's Rings in Motion

First spied in 1619 by Galileo, who described them as "handles," Saturn's rings remained a mystery until 1655, when the Dutch scientist Christian Huygens realized they were a system of rings around the planet. More than 300 years later, they were unveiled in all their individuality by the *Voyager* space probes in the 1980s. In the photo at left, the shadows of the A, B, and C rings mark the almost featureless surface of Saturn, shown with the tiny moons Tethys and Dione. The false-color image at right reveals more than 60 bright and dark ringlets in the C ring and suggests the differing surface composition of the particles making them up. *Voyager's* data also gave scientists clues to the gravitational forces (*below*) that maintain the narrow bands, spiral patterns, and twisted braids that characterize the rings.

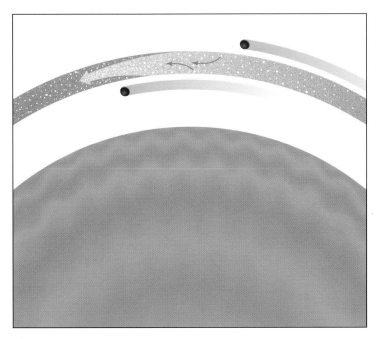

Shepherd Moons

In this greatly exaggerated view, looking directly down on the ring plane, we see the actions of two moonlets on ring particles. A small outer moon drags on particles drifting outward, robbing them of orbital energy so that they drop to lower orbits. The inner moon, traveling faster than particles farther out, tugs the particles in their direction of motion. This transfers angular momentum and boosts the slower-moving particles into higher orbits. The one-two action of the shepherds keeps the particles confined to a narrow band.

Gravitational Waves

In this simplified cross section of a ring, spiral density waves alternately bunch particles and disperse them (*near right, center, and far right*). Also shown is a bending wave, a phenomenon triggered by interaction with an orbiting moon that pushes the ring sheet into crests and troughs.

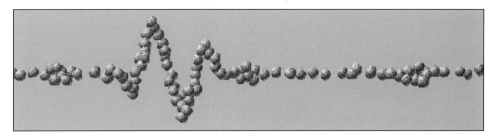

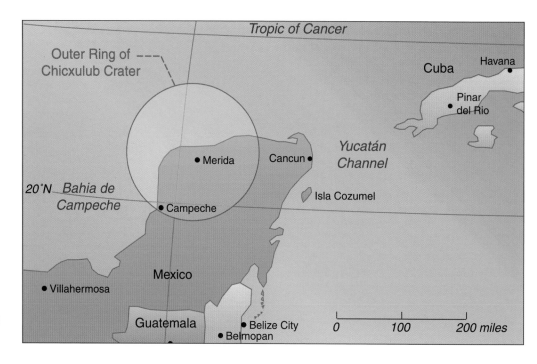

Formed by the impact of an asteroid or a small comet that smashed into Earth about 65 million years ago, the Chicxulub Crater in Mexico's Yucatán Peninsula may measure nearly 200 miles across. An impact of this magnitude would have created huge ocean waves and a global dust cloud that blocked sunlight for years.

approach it closely. Hundreds more are likely to exist. Finding them is no easy task because they are faint and move quickly across the sky. Nevertheless, astronomers hope to spot most of the truly dangerous ones—those measuring a half mile across or more—by early in the twenty-first century.

If you think that's frivolous, you need look no further than Jupiter itself. In 1993 the planetary scientists Eugene and Carolyn Shoemaker and the astronomer David Levy discovered a comet that looked strangely flattened. They and their colleagues soon realized that the comet, named Shoemaker-Levy 9, was trapped by Jupiter's gravitational field. The planet's intense tidal forces had torn the comet into 21 pieces, each about a mile wide, and stretched them into a 100,000-mile-long chain that looked like a string of pearls (page 58). Orbital calculations revealed that these pieces would plow directly into Jupiter on their next pass by the planet. Astronomers eagerly set up their telescopes for the first recorded collision of objects in the solar system.

The results were breathtaking. By July 1994 the chain was over a million miles long. The fragments took a week to slam into the planet, one after the other. Several of them carved dark, long-lived scars in the atmosphere while blasting plumes of gas thousands of miles into space. The seventh fragment exploded with an energy equivalent to 6 million megatons of TNT—nearly 1,000 times the power of all the nuclear weapons on Earth—and created a fireball 2,000 miles high. If any of the 12 largest pieces had struck Earth, the human race probably would have been obliterated.

A blast with the force of a 10-megaton bomb flattened trees in the Tunguska region of Siberia in 1908. Scientists think the destroyer came from space, probably a small comet that detonated in midair.

In one of the ironies of nature, Jupiter's disruptive presence in our solar system is also a blessing. Without Jupiter's tendency to sweep aside intruders such as incoming comets, Earth almost certainly would be more heavily bombarded. However, an asteroid did sneak through 65 million years ago, gouging a deep hole near Mexico's Yucatán Peninsula and probably dooming the dinosaurs. A mile-wide crater in the Arizona desert bears witness to an impact 50,000 years ago by a rock as wide as a football field. And just the other day, in cosmic terms, a vast stretch of forest near the Tunguska River in Siberia was flattened in 1908 when some object—probably a small comet—detonated in the air with the force of a 10-megaton bomb. These events serve as reminders that motion in our seemingly tranquil universe can unleash great violence with little warning.

Gravity Rules

Like a traffic cop in the middle of a busy intersection, Jupiter directs the comings and goings of the asteroids and comets that whiz around the solar system. Or rather, Jupiter's gravity directs traffic. According to the field of science known as chaos theory, asteroids in particular orbits around the Sun tend to have chaotic gravitational interactions with the giant planet. Their relationships are so sensitive that the slightest push or pull or gravitational brush with another interplanetary object can have unpredictable long-term consequences.

In some senses, Jupiter's massive presence protects Earth from what would probably be much more frequent bombardment by wandering comets and by the asteroids that inhabit the zone between Mars and Jupiter. However, over the course of millions of years, Jupiter's relentless gravitational tug has yanked some of those asteroids out of the equatorial plane, sending them into highly elliptical paths that cross the orbit of Mars. Gravitational interactions with the Red Planet further perturb the asteroids into even more eccentric orbits that cross paths with Earth and Venus. Indeed, astronomical observations reveal that Earth is surrounded by a swarm of Earth-crossing asteroids and comets (*opposite*). Bombardment by these missiles was and continues to be part of Earth's geological process.

Even Jupiter is not immune. As the photos on the following pages show, the tidal forces exerted by the gas giant might shatter an incoming asteroid or comet into a score of pieces, but the planet itself cannot escape the inexorable rule of gravity.

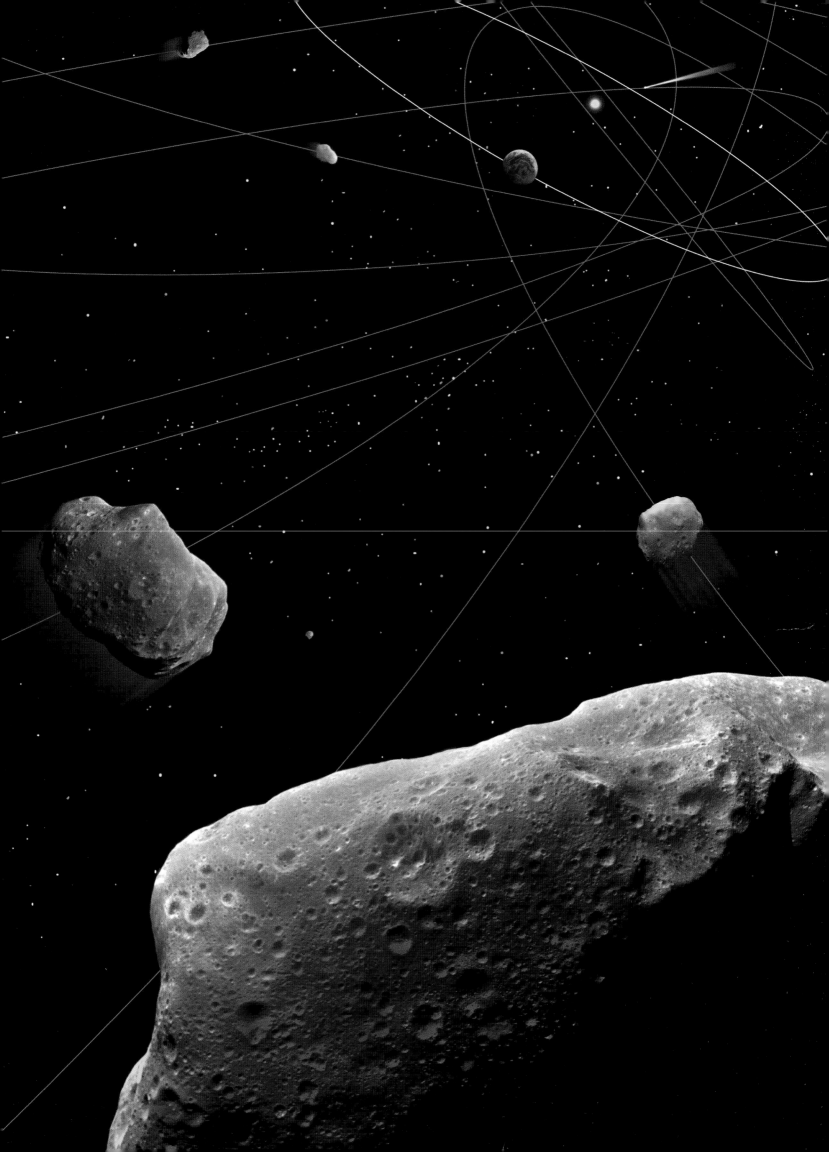

Encounter with a String of Pearls

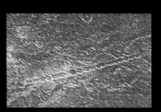

For six days in July 1994, fiery fragments of Comet Shoemaker-Levy 9 rained down on Jupiter, the first collision of two solar system bodies ever witnessed. Measuring as much as one and a quarter miles wide, the fragments slammed into Jupiter's atmosphere at more than 30 miles per second. The fragments all struck Jupiter from the south as the planet rotated beneath them, leaving scars in the atmosphere that lasted for weeks. Had the fragments struck a solid surface, Jupiter would be sporting a vast necklace of craters, an extended version of a chain of craters on its moon Callisto (*above*). At one impact site in the Jovian atmosphere, one of the fragments left a dark thick ring the size of Earth (*opposite*).

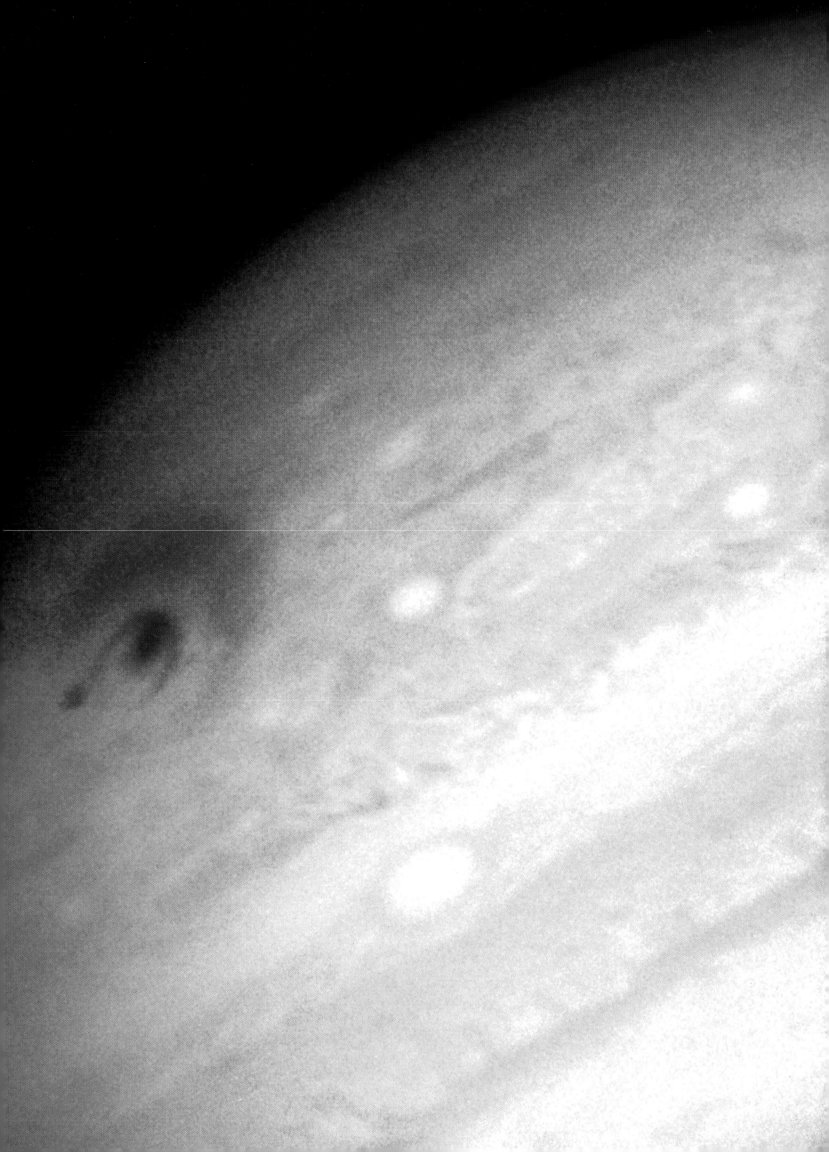

Matter

The Stuff of the Universe

Glowing in the morning sky, Comet Hale-Bopp shows the distinctive tail that forms as ice in the comet's nucleus sublimes, or transforms directly from solid to gas—the same process that makes ice cubes mysteriously shrink in the freezer.

A bright visitor blazed through our solar system more than four millennia ago as Egyptian laborers built the great pyramids. The laborers may have gazed at this "hairy star," wondering at its swift pace through the heavens and its shifting shape from night to night. When this same visitor returned in the late twentieth century from a cold hibernation far beyond the orbit of Pluto, we called it Comet Hale-Bopp. The comet graced the sky for months, easily visible without a telescope even from light-polluted cityscapes. In the countryside it was a breathtaking sight, with a gauzy tail stretching millions of miles through interplanetary space.

More people may have seen Comet Hale-Bopp than any other comet in history. Those who followed its journey closely were treated to a vivid illustration of the ever-changing nature of matter in our universe. Since its previous passage near Earth in the twenty-third century B.C., the comet had spent most of its time drifting as a quiet lump on a cigar-shaped orbit 50 billion miles long. Light from the distant Sun was too feeble to heat the comet's skin. Its ices and dust were locked in a frozen embrace, relics of the interstellar material that formed our solar system nearly 5 billion years ago.

As the comet neared the inner planets, the Sun's gravity tugged it inward at a quickening pace. The growing warmth transformed the comet's surface. The outer-most layers—consisting mostly of frozen water, carbon dioxide, methane, and ammonia—started to evaporate into space. This made the comet shrink, by the same process that withers ice cubes to sour-tasting nubs when you leave them in your freezer too long. (In this process, called "sublimation," solids transform directly into gas without passing through a liquid state.) Gaseous geysers erupted at weak spots on the surface, expelling dust and spewing vaporized ices into a puffy cloud, or coma, around the comet's nucleus. The dust came from impurities in the comet's ices that were left behind when the ice sublimated. This coma reflected enough sunlight to become visible through telescopes on Earth when the comet was still more than 500 million miles away, beyond the orbit of Jupiter. On the evening of July 22, 1995, the professional astronomer Alan Hale and amateur astronomer Thomas Bopp independently spotted the slowly moving fuzzy blob within minutes of each other. Their discovery gave the comet its catchy hyphenated name.

Comet Hale-Bopp's coma grew larger as the comet fell toward the Sun. Gradually, the combined influence of sunlight and solar wind—a steady blast of charged particles that blow outward through the solar system—pushed the dust and gas into a stream pointing away from the Sun. This tail, the classic signature of a comet, grew as long as 50 million miles when Comet Hale-Bopp ventured closest to the Sun in April 1997. Careful viewers may have noted that the comet displayed two distinct tails. Most obvious was the bright dust tail, illuminated by the Sun's rays as it swept away from the comet's nucleus like a witch's broom. The gentle pressure of sunlight forced these tiny grains outward along curved paths as the comet moved through space. In dark skies a second tail also appeared: an ethereal gas tail, shining with a faint blue light emitted by the gas particles themselves. The electrically charged solar wind propelled this gas on a straight path away from the Sun.

On certain nights you may have seen stars shining through Comet Hale-Bopp's tail as if it wasn't there at all. Indeed, comet tails are among the wispiest collections of matter imaginable. The puffs of gas and dust stretch into a huge volume of space, just as a small cloud in the sky holds only a cupful of water sprayed into tiny droplets of vapor. Reflected sunlight makes both the cloud and the comet tail appear much more substantial than they really are. If we could have compressed Comet Hale-Bopp's tail to the density of the air we breathe, its 50-million-mile length would have fit into a cube no more than 10 miles on a side. The Harvard astronomer Fred Whipple, the first to characterize comets as "dirty snowballs," succinctly described a comet's tail as "the most that has ever been made of the least."

The ancient Romans believed that comets augured the births, deaths, and fortunes of kings. A coin commemorating Julius Caesar (*above*) shows on its reverse side an eight-pointed comet-star that was seen in broad daylight in 44 B.C. and was taken to be the soul of the assassinated emperor. In another example of the seeming association of the appearance of comets and the fate of kings, a detail of the Bayeux tapestry (*left*) displays a comet—now known to be Halley's—which in A.D. 1066 was thought to foretell the death of England's King Harold at the Battle of Hastings.

Tails of a Comet

As seen in this photo of Comet Hale-Bopp, comets often develop two types of tails as they round the Sun at perihelion, their closest approach. Stars can be seen through the comet's bright blue ion tail, which is starting to develop some streamers. The yellowish dust tail is equally ethereal. Both tails point away from the Sun, orienting themselves according to where the comet is in relation to the Sun and the charged particles of the solar wind (*opposite*).

The Sun's strong gravitational pull dooms comets to return again and again to the inner solar system. Each time they lose more mass and shine a bit less brightly. Some comets fizzle after hundreds or thousands of such orbits and drift like small asteroids among the inner planets. Others split into pieces like poorly packed snowballs if they wander too close to the Sun. Still others plunge into the Sun and melt in a flash—the ultimate in solar system recycling.

As for Comet Hale-Bopp, our descendants will see it again in the year 4380. Its return trip will be much faster, thanks to a gravitational tweak in its orbit when the comet passed Jupiter in 1995. Between now and then hundreds of other bright comets will streak across the sky. Each will remind its admirers of how easily matter can transform from ancient ice to a mist that vanishes into space on the breath of the solar wind.

MATTER'S Many Guises

Comets are beautiful examples of a rule of thumb that applies to most matter in the universe: Sooner or later it changes into something else. Consider the wooden furniture in your home, which seems permanent enough. But not long ago the furniture was a tree, and before that it was a seedling that grew by using sunlight to convert water, carbon

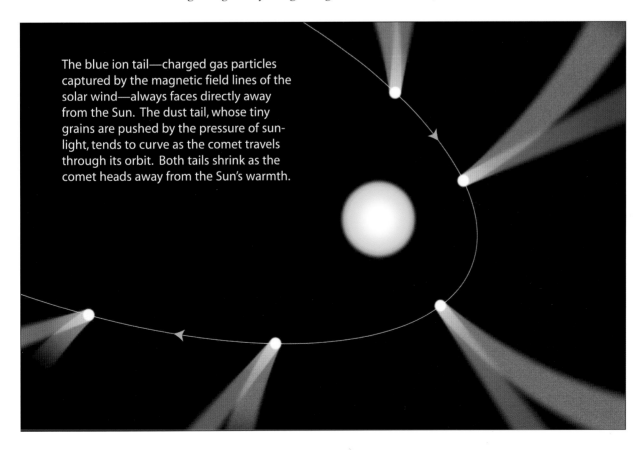

The blue ion tail—charged gas particles captured by the magnetic field lines of the solar wind—always faces directly away from the Sun. The dust tail, whose tiny grains are pushed by the pressure of sunlight, tends to curve as the comet travels through its orbit. Both tails shrink as the comet heads away from the Sun's warmth.

dioxide, and nutrients into new wood. Not long in the future the furniture will rot and return to the soil, or perhaps burn and waft into the atmosphere as soot. Earth will out-live these belongings, but 5 billion years from now it too will incinerate as the Sun runs out of fuel and bloats into a giant ball that engulfs the inner planets. The stars them-selves live and die, converting matter into energy along the way. Their deaths seed space with fresh bursts of matter—the raw material needed for the next generation of stars.

When matter undergoes a dramatic change, we usually can trace the transforma-tion to changes in temperature, pressure, or density. On Earth we are accustomed to a narrow range of these properties. For instance, we live mostly at room temperature. Our notions of "hot" and "cold" span just a few dozen degrees on the thermometer. We have also adapted to the pressure of air and water near sea level. When we fly at high altitudes or dive deep into the sea, we need to enclose ourselves in vessels with similar pressures for our bodies to function. Similarly, we experience gases with densities typical of air, liquids with densities close to that of water, and solids with densities like the ground under our feet.

These natural biases make conditions elsewhere in the universe seem exotic. But in fact our Earthly conditions are rare. Temperatures in space plummet to near absolute zero (minus 273 degrees Celsius), the point at which all motion ceases except for ever-present quantum vibrations. Things are more extreme on the hot end of the scale, where temperatures soar to tens of millions of degrees in the cores of stars. Our sister planet Venus has an atmospheric pressure nearly 100 times greater than Earth's, enough to squash a nuclear submarine. The thick atmospheres of the planet Jupiter and the other gas giants create pressures thousands of times higher still. Deep inside the planets, this overwhelming force converts gases into bizarre states with liquid or even metallic properties. The range of densities in our universe is staggeringly large as well, from the emptiness of intergalactic space to the crushed interiors of dead stars.

We can gain some sense of how matter responds to these alien settings by re-creating extreme conditions in laboratories on Earth. For example, modern vacuum pumps, such as those used to insulate cryogenic components or to evacuate particle accelerators for astronomy and physics research, easily remove 99.99999 percent of all air from a sealed container. The best scientific pumps in the world can make small vacuums thousands of times better than that. Experiments in such vacuums offer clues about how matter may behave in the cold emptiness of space, where interactions between atoms and molecules are rare.

Even so, our vacuum chambers are crude approximations of the cosmos. Concentrations of gas between stars, such as the colorful Orion Nebula, contain up to a billion atoms and molecules of gas per cubic yard of space. Amazingly, that is 10 million times less dense than the best laboratory vacuums ever produced. Most of the rest of our galaxy is a thousand times more rarefied still. The space between galaxies is the most desolate void of all: less than one atom drifting in each cubic yard. Imagine a volume of space measuring 125,000 miles on a side—a box big enough to stretch halfway to the Moon. In intergalactic space such a box would contain about as many atoms as the air in your refrigerator.

Here's another way to think about the rarity of matter. If we spread all of the universe's matter evenly throughout space, we would see just a few atoms in each cubic yard. To achieve that same density with the number of atoms in half a grain of rice, we'd have to expand the grain to the size of Earth. Needless to say, our local environment is one of the rare spots in space that contain matter in abundance. An average cubic yard of air at Earth's surface holds about 50 trillion trillion atoms, making our planet a place where matter matters.

Other clever tools help us explore the realm of hot and dense matter. One such apparatus is the diamond anvil cell, a vise equipped with two flat-tipped diamonds. Physicists place tiny mineral samples between the diamonds, crush them at pressures mimicking those near Earth's core, and heat them with lasers to thousands of degrees. This opens an experimental window on the interior of our planet, which otherwise is shielded by thousands of miles of rock. The results help us understand how minerals behave at key boundaries within the planet. For example, Earth's solid rocky mantle lies atop a core of liquid iron, with an abrupt and active transition between these layers. At the center of that molten core sits a ball of solid iron. Geophysicists can determine its depth with laboratory experiments that gauge how iron reacts to the hellish temperatures and pressures of inner Earth. Other rocky planets and moons appear to have similar internal layers, although only Jupiter's moon Io is as dynamic as Earth.

To delve into even more extreme conditions, we must call upon the most powerful lasers yet invented. Physicists use pulses of light that carry as much energy as the entire

Imagine a volume of space measuring 125,000 miles on a side—a box big enough to stretch halfway to the Moon. In intergalactic space such a box would contain about as many atoms as the air in your refrigerator.

electrical grid of the United States, but only for a trillionth of a second or so. These beams eradicate small targets placed in their paths. The temperatures and pressures within the tiny blasts approach those inside our Sun or planets like Jupiter. One of the goals of this research is to harness the energy of thermonuclear fusion that powers the Sun. That would be a much cleaner source of energy than nuclear power from the fission of uranium. However, it may take decades to learn how to sustain the fierce fusion reactions in a controlled and profitable way.

In the meantime, the experiments have shown us that hydrogen—the main component of Jupiter, Saturn, Uranus, and Neptune—takes on distinctly ungaslike properties as pressures and temperatures rise within those planets. For example, lasers have compressed and heated hydrogen into a form that appears to conduct electricity as efficiently as a metal. This odd transformation may in fact occur near Jupiter's core, helping to produce a powerful magnetic field around the planet.

We clearly still have much to learn about how matter behaves as we move from one extreme in the universe to the other. Even so, our understanding of the nature of matter has evolved considerably since the Greek philosopher Leucippus and his student Democritus first proposed the idea of the atom in about 440 B.C. Leucippus and Democritus pondered how long a piece of iron would retain the basic properties of iron if one broke it in half again and again. They theorized that there was a basic particle, a corpuscle of matter, beyond which one could go no smaller. All matter in the universe, they reasoned, was made of these "atoms," from the Greek word for "indivisible."

Not until the early twentieth century did we learn that atoms weren't simply an idea of convenience. The New Zealand–born physicist Ernest Rutherford did the most to prove their existence and discern their structure. Prior to his work, physicists envisioned atoms as diffuse blobs. This model held that negatively charged electrons were embedded in the blobs like raisins in a positively charged plum pudding. Rutherford and his colleagues tested that notion by firing particles at a thin gold foil. The particles, which themselves carried a positive charge, came from the radioactive decay of a small amount of uranium and moved at 5 percent of the speed of light. Most particles streamed through the gold foil, as expected. However, a tiny fraction bounced at sharp angles or even reflected back toward the gun. This result amazed Rutherford. As he said later, "It was almost as incredible as if you fired a 15-inch shell at a piece of tissue paper and it came back and hit you."

It immediately became clear that the "plum pudding" model of the atom didn't work. A diffuse spread of positive charge within the atoms couldn't possibly make any incoming particles ricochet backward. Instead, it seemed, the particles encountered hard nuggets of positive charge in the atoms and were repelled, just as the north poles of two magnets repel each other. When Rutherford calculated how concentrated those charges had to be, he determined that each gold atom contained a nucleus measuring just 1/100,000th the diameter—and 1 million-billionth the volume—of the entire atom. The electrons darting around this nucleus carried the atom's negative charge but virtually no mass. It was no exaggeration for Rutherford to conclude that atoms were almost entirely empty space.

To put his shock into visual terms, imagine enlarging an atom until its cloud of electrons fills the volume of the Louisiana Superdome in New Orleans. The site of many football Super Bowls, the Superdome is nearly 700 feet across. Now picture a single ball bearing, one-twelfth of an inch wide, suspended in the center of the dome's cavernous volume. The ball bearing represents the atom's nucleus, surrounded by electrons flitting about within an enormous void. That's an accurate scale model of an atom as implied by Rutherford's work. Indeed, if we could somehow remove the spaces from within the atoms that make up our planet, the entire Earth would fit easily under the Superdome's roof.

Given the seemingly porous nature of every atom on Earth, what keeps us from walking through walls or sinking into the ground? The answer is electrostatic repulsion on an atomic scale. Clouds of electrons around every nucleus create what amounts to a ball of negative charge. Since negative charges repel each other, the electron clouds set up impenetrable force fields around atoms. Only the catastrophic crushing power of a dying star can overcome that barrier (page 93). Two other forces operate on the tiny scale of a nucleus. One is the aptly named strong nuclear force, which binds together protons (positive charges) and neutrons (neutral or no charge) within the nucleus. The second is the weak nuclear force, which mediates the radioactive decay of unstable elements. Don't let the name fool you, however, because the weak nuclear force is still vastly stronger than gravity on these minuscule scales.

Physicists probe the precise workings of these forces by smashing together particles in powerful accelerators. Their results apply not only to matter on Earth but

"It was almost as incredible as if you fired a 15-inch shell at a piece of tissue paper and it came back and hit you."

Discovering the Atomic Nucleus

In 1909 the physicist Ernest Rutherford directed an experiment at the University of Manchester in England to measure small deflection angles recently observed when alpha particles—tiny positively charged bodies given off by radioactive elements—were beamed through a thin gold foil. He wanted to gauge the distribution and charge of matter within the atom. The then-current theory held that the atom was like a pudding, a diffuse, positively charged sphere studded with negatively charged electrons.

As illustrated here, the setup included a radioactive source, a target consisting of a thin sheet of gold foil, and a detector consisting of a screen covered with zinc sulfide. Although atoms and subatomic particles are much too small to be seen directly, particles hitting the screen would leave microscopic marks in the zinc sulfide. The pattern of marks in the screen (*opposite*) was a huge surprise and could not be accounted for by the "pudding" atom. In 1911 Rutherford proposed an explanation: The atom was largely empty space, so most of the tiny alpha particles could pass unimpeded. But in the center of the atom was a minuscule and highly charged nucleus that held most of the atomic mass—something like the Sun in the center of the solar system. Alpha particles deflected by the nucleus bounced back in unexpected ways.

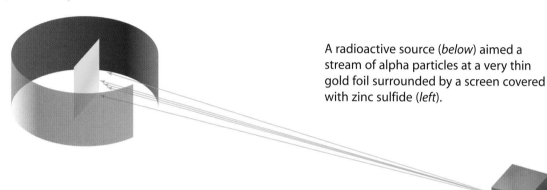

A radioactive source (*below*) aimed a stream of alpha particles at a very thin gold foil surrounded by a screen covered with zinc sulfide (*left*).

also to distant planets, stars, and galaxies. Consequently, physics experiments have led to a "standard model" for the nature of matter and forces throughout the universe. The basic constituents of matter in this standard model are not protons or neutrons, but quarks (page 75), the fancifully named specks first proposed in 1964 by the American physicists Murray Gell-Mann and George Zweig. In the end, quarks may not prove to be nature's ultimate building blocks, but for now they explain all the properties of matter we have observed. Most physicists think that an overarching theory will supersede the standard model, just as relativity superseded Newton's laws of motion. Such a theory would combine all known forces and particles into one elegant recipe of the cosmos.

Any good theory of nature must account for the properties of elements, the fundamental substances that cannot be broken down into something simpler. The ancient Greeks thought there were four elements: fire, air, water, and earth. They

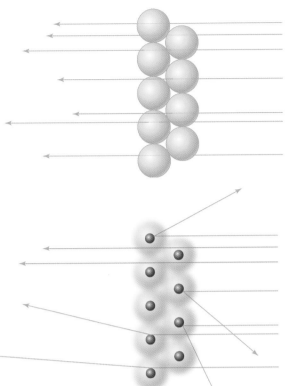

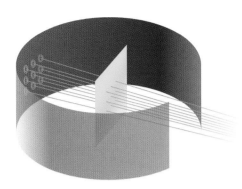

Expected Results

High-energy alpha particles should pass through a thin gold foil only a few atoms thick (*left*), leaving a small region at the back of the zinc screen covered with dots (*right*).

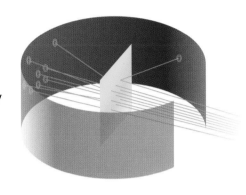

Unexpected Results

As predicted, dots appeared mostly at the back of the screen (*right*), but every so often dots were scattered near the front of the screen, as if they had ricocheted.

New Atomic Theory

Rutherford's explanation was that the ricocheting alpha particles had bounced off something small, dense, and positively charged in the center of the atom (*above*). The new model of the atom—largely empty space with a compact nucleus—both explained the surprising results of the experiment and was a milestone in modern atomic theory.

also proposed a fifth element: the essence of the heavens, called "quintessence." Today, we know of more than 115 elements, about 90 of which exist naturally. (Scientists forge the rest in their labs, but most of them exist only for fractions of a second before they decay radioactively into other elements.) Think back to your childhood, when you may have played with Tinkertoys or other sets of interconnecting pieces. You created a seemingly endless array of contraptions using perhaps a dozen different shapes and sizes of building blocks. Now, imagine such a game with 90 varieties of pieces. That's the flexibility nature has at its disposal to construct our universe.

However, nature cannot mix elements in a haphazard way. Certain rules dictate which elements can combine. For instance, "noble gases" such as helium and neon almost never react with anything, just as a noble lord might ignore the common folk in his domain. Fluorine and chlorine, on the other hand, grab hold of just about anything that comes along. A well-known chart called the periodic table of the elements helps students of chemistry figure out which elements can react, how many atoms each element must contribute to do so, and what the properties of the resulting compound will be. The rows and columns of the periodic table are not especially artistic. Rather, its patterns—the periodicity of its design—nicely illustrate the origins of chemical behaviors on Earth and in the cosmos that otherwise might seem random.

To understand why the periodic table works so well, we must turn again to quantum mechanics. Newton's laws of motion don't work well on atoms. Electrons in particular pose a special problem. They are so tiny and move so fast that classical physics fails utterly in describing them. Many physicists helped explain this failure by devising the tenets of quantum mechanics in the early twentieth century. But the one rule that explained the periodic table most convincingly was the Pauli exclusion principle, named for the Austrian physicist Wolfgang Pauli. This rule states that no two electrons can inhabit the same space at the same time and move in exactly the same way. Instead, electrons fill a series of "shells" around the nucleus of an atom. If one of an atom's shells isn't completely full, the atom can either donate electrons to other atoms or accept electrons from them. The combined atoms then form a stable molecule. The configuration of the periodic table shows exactly which such bonds can occur. We can

Clouds of Probability

According to quantum theory, electrons orbit the nucleus of an atom in a series of orbital shells that correspond to the energy level of the electron. No two electrons of exactly the same properties can inhabit the same space at the same time. Thus, atomic orbitals are characterized as "probability clouds"—the volume of space around the nucleus of an atom in which an electron of a given energy has a 90 percent probability of being found. A given orbital can be described by quantum numbers pertaining to the size of the atom and the general shape and orientation of the orbital. Because different elements have different numbers of electrons, their orbitals take on a variety of possible shapes and orientations. Shown here are three of the many possible orbital configurations.

The lowest energy state produces a simple *s* orbital (*left*). A *p* orbital (*above right*) has two lobes separated by a plane, where the probability of finding an electron is zero. One of five possible orientations of a four-lobed *d* orbital is shown at right.

The Periodic Table

By the late nineteenth century chemists realized that if they arranged the elements in increasing order of atomic weight, the chemical properties of the elements would then repeat in a periodic manner. If the elements were listed horizontally by increasing atomic weight, groups of elements with similar chemical properties fell in vertical columns. For example, the alkali metals lithium, sodium, and potassium (*below left*) have similar properties as do the halogens fluorine, chlorine, bromine, and iodine (*below right*). The scheme allowed for the prediction of elements not yet discovered.

A better understanding of why certain elements share certain properties came with the advent of quantum mechanics. Atoms of different elements contain different numbers of electrons. Hydrogen, for example, has just one electron, helium has two, and carbon has six. These fast-moving subatomic particles fill the atom's orbital shells to different degrees. With one electron the hydrogen atom has a vacancy in the first orbital shell; it therefore combines easily with itself and with other elements. By contrast, helium's two electrons fill its first orbital shell; like all the noble gases, it remains inert and nonreactive.

The modern periodic table arranges the 110 known elements in order of increasing atomic number, which is the number of protons in the nucleus and also therefore the number of electrons surrounding the nucleus. Hydrogen, with one proton and one electron, is at top left. Helium, with two protons, is at top right. These two elements make up the first period, with other elements in rows 2 through 7 listed in increasing order of atomic weight. As seen below, element 110 has a temporary designation. It and several other more recently discovered elements (111, 112, 114, 116, and 118) await official nomenclature.

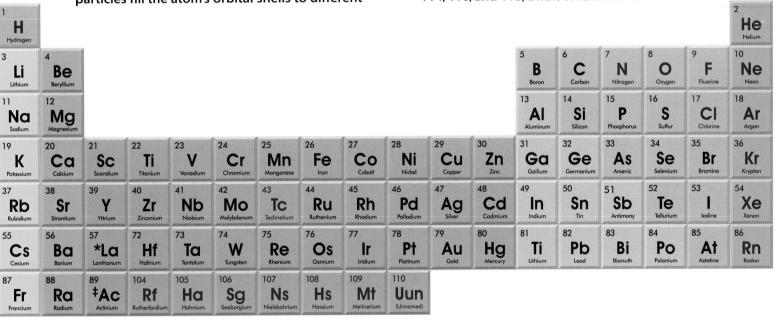

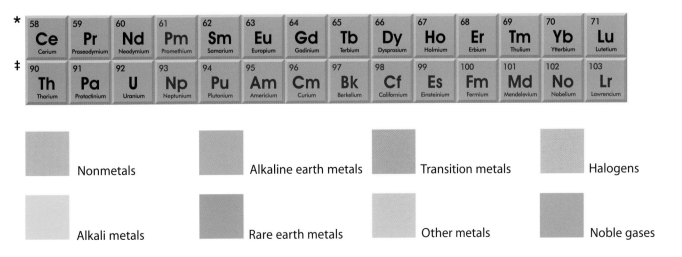

apply those rules to a host of chemical processes on Earth, such as predicting the noxious compounds that form when car exhaust drifts into the atmosphere. The rules work equally well in space. For instance, astrophysicists can determine which elements combine in cooling interstellar gas clouds. The new molecules created in this way eventually lead to new stars and planetary systems.

The periodic table is so basic to understanding the chemistry of the cosmos that a panel of physicists, archeologists, artists, and sociologists recently devised a surprising way to use it. The panel was charged with creating warning systems for the Waste Isolation Pilot Plant (WIPP), an underground storage facility in New Mexico for low-level radioactive waste. The experts had to envision systems that would warn people against digging at the WIPP site for at least 10,000 years because its contents will stay hazardous for that long. Languages and cultures are likely to come and go in that time, but WIPP will remain. Panel members designed frightening sculptures, earthworks, and other symbols of danger. They also proposed a chamber containing an engraved reproduction of the periodic table, with highlights marking the squares for uranium, plutonium, and other radioactive elements. Any future scientists would recognize the hazard, the panelists reasoned, because the periodic table is likely to endure.

The Scarcity of MATTER

The periodic table is a tool for us to understand how matter behaves, and Rutherford's model of the atom helps us realize that all matter is mostly empty space. We also have seen that matter is rare in the universe—just a few atoms per cubic yard of space, on average. But by a different reckoning, it seems there is plenty of matter to go around. There are perhaps 100 billion galaxies in the universe, each containing perhaps 100 billion stars. Every person on Earth would have to count five stars per second for about 10,000 years to tally all of those stars, not to mention the atoms that compose them. How is it possible for so much matter to add up to so little?

The key is to grasp the vast distances between objects in the universe. Just as we constructed a model of an atom with a tiny ball bearing in the center of the Louisiana Superdome, we can imagine scale models of planets and moons in our solar system, stars in our galaxy, and groups of galaxies in the cosmos as a whole.

Subatomic Particles

By smashing atoms at ever-higher energies to produce increasingly exotic subatomic particles, physicists have learned that most of the tiny specks are actually combinations of a small number of fundamental objects. According to current theory, the two most fundamental subclasses of particles are fermions, named for the Nobel laureate Enrico Fermi, and bosons, named for the Indian physicist Satyendra Bose. Fancifully illustrated here are the fermion members of the subatomic family along with their antimatter counterparts. Bosons, which transmit the four known forces between fermions, are shown on page 76.

The most fundamental fermions are classified as either leptons or quarks. Quarks are bound together in threes by the strong nuclear force to make neutrons and protons. Less common particles such as pions and kaons are made up of two quarks. Experiments suggest that quarks come in six varieties, which physicists have named up, down, charm, strange, top, and bottom.

Leptons, which are all low-mass objects, do not combine with each other or with other particles except under special circumstances. They come in six varieties: Three have a negative charge and three have no charge. Charged leptons include electrons, muons, and taus. The heaviest is the tau, with nearly twice the mass of a hydrogen atom. Leptons with no charge are called neutrinos ("little neutral one" in Italian) and usually accompany a charged counterpart. Thus, they include the electron-neutrino, the muon-neutrino, and the tau-neutrino. Once presumed to be massless, neutrinos are now suspected of possessing just a smidgeon of mass.

A generic fermion is shown here with its antiparticle, which has the same mass, although all other properties are reversed.

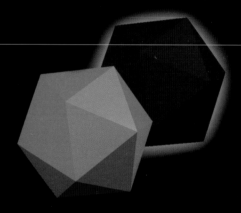

The electron, carrier of electric current and of negative charge in the atom, has as its antiparticle the positively charged positron.

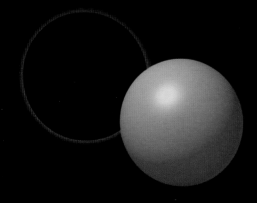

Generic quarks and antiquarks are bound by the strong nuclear force to form composite particles.

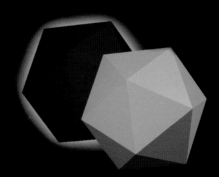

The neutrino and antineutrino have no charge and are believed to have minuscule mass.

Bosons: Particles That Carry Force

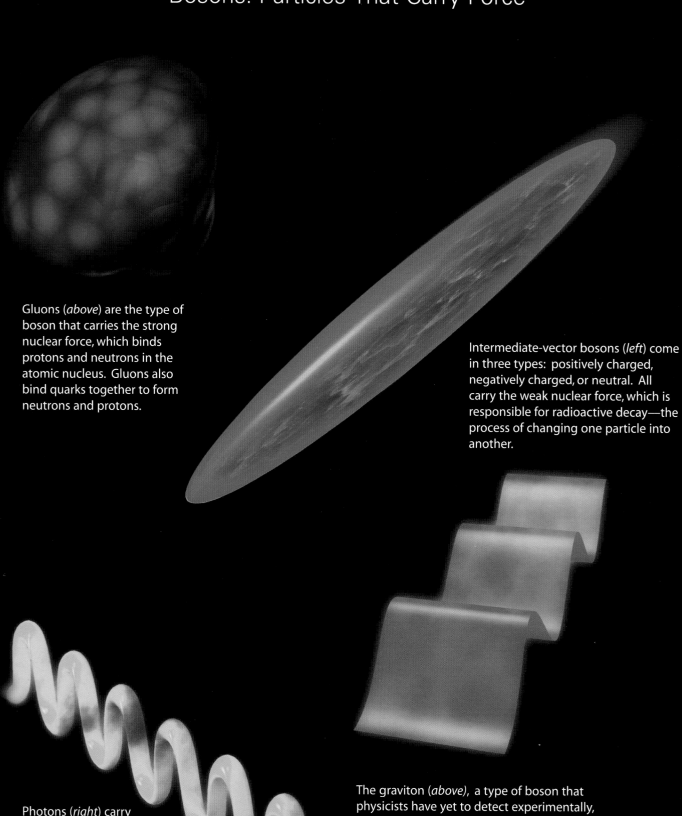

Gluons (*above*) are the type of boson that carries the strong nuclear force, which binds protons and neutrons in the atomic nucleus. Gluons also bind quarks together to form neutrons and protons.

Intermediate-vector bosons (*left*) come in three types: positively charged, negatively charged, or neutral. All carry the weak nuclear force, which is responsible for radioactive decay—the process of changing one particle into another.

Photons (*right*) carry electromagnetism between charged fermions such as electrons.

The graviton (*above*), a type of boson that physicists have yet to detect experimentally, is believed to carry gravity among all particles in the cosmos.

By comparing celestial objects with everyday things, the unimaginable distances in the cosmos grow slightly more tangible.

For starters, consider our Earth and Moon. Our companion in space has one-eightieth the mass of Earth, making the Earth–Moon system verge on being a double planet. If we shrank the system so that Earth was the size of a basketball, the Moon would be the size of a softball about 30 feet away. That distance corresponds to the farthest we have ever sent people into space. On this same scale, Mars (at its closest) is a mile away from Earth. Making the leap to send astronauts to Mars may seem like a simple extension of the Apollo missions, but hopping 30 feet is nothing at all like hopping a mile.

The basketball analogy also is useful for picturing the scale of our solar system. But this time the basketball-sized object will be our Sun. We'll mix our sports analogies just a bit and put the basketball-Sun at home plate on a baseball diamond. On that scale the innermost planet, Mercury, would look like a pellet of birdseed orbiting at a distance of 35 feet—halfway to the pitcher's mound. Earth becomes the size of a sunflower seed about 90 feet away, the distance to first base. Jupiter is a one-inch marble 450 feet away, just over the fence in center field. And tiny Pluto, a grain of pepper, orbits in isolation about 3,500 feet away, far beyond the parking lot. Throw in five more planets and scatter some sand to represent asteroids and you have our model solar system: a sphere more than a mile across containing a basketball at its center, flecks of matter here and there, and lots of empty space.

A vast swarm of trillions of comets, called the Oort Cloud, probably surrounds the solar system as well. The comets drift in cold storage in the depths of space, perhaps as far away as 1,000 times the distance of Neptune from the Sun. In our model that's like dust motes in Baltimore orbiting a basketball-Sun in New York. The bond keeping the comets in orbit around the Sun is exceedingly weak, but it's a testament to the long-range strength of gravity.

The true eye-opener comes when we consider the distances between stars. Our basketball-Sun in New York would have as its nearest neighbor another basketball in Honolulu—5,000 miles away. Except for a few clouds of gas, the space in between is largely devoid of matter. Hollywood movies often show starships cruising through the galaxy, drifting past stars like fireflies at the rate of one or two every second. But the gaps between stars in the galaxy are so great that the vessels would have to travel up to 500 million times faster than the speed of light to pass stars so quickly. We doubt that Scotty could supply the necessary power from his warp engines on the *Enterprise*, even in his finest moments.

Other movies portray galaxies as swirls of light within which heroes and villains dash around at will, from one side to the other. To see how realistic that is, let's extend our basketball-Sun model to our galaxy as a whole. If the Milky Way's average stars were the size of basketballs, the galaxy would measure an astonishing 125 million miles across. That's an enormous volume in which to spread out 100 billion basketball-suns. Since it's also too big for most of us to grasp, we'll shrink the scale of our model much further. Imagine squeezing our entire solar system—planetary orbits and all—into a coffee cup. (The Sun would be the size of the tip of a pin in the center of the cup.) On that scale the Milky Way still would cover an area as large as North America.

To consider the distances between galaxies, let's take one more step and collapse the entire Milky Way into that coffee cup. The next nearest large galaxy—called Andromeda—then would hang in space about 7 feet away. Compared to the vast separations between individual stars, galaxies actually are fairly close neighbors in space. Surveys of chunks of the universe have revealed that galaxies are arrayed in clumps and spidery filaments, congregating around gaping voids that contain virtually no galaxies. Data suggest that these patterns trace the imprint of microscopic irregularities in the structure of space-time that arose during the first instants of the Big Bang.

Our series of earthly analogies may help your mind leap from one scale of the cosmos to the next. However, such descriptions do not capture the majesty of matter that inhabits these vast spaces. In that way, pictures truly are worth more than the words we use to describe them. Several images from recent decades have raised our awareness of our place in the cosmos. One such picture came from the *Galileo* spacecraft. On its way to Jupiter, *Galileo* looked back from beyond the Moon at the Earth–Moon system. The fragile blue crystal of Earth and the familiar crescent of the Moon formed a cozy pair in space. The small gap between them made it clear

A rare perspective, photographed by the *Galileo* spacecraft in December 1992, captures the close relationship of Earth and its only natural satellite.

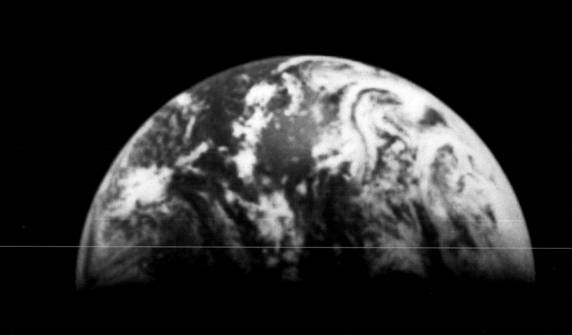

that the Apollo astronauts did not venture far from home at all—and that our planet is the human race's only home for the foreseeable future.

Another stirring image came from the Hubble Space Telescope, which stared at a single tiny patch of sky for days on end. The telescope actually captured two such "deep-field" images, one in Northern Hemisphere skies and one in the Southern Hemisphere. They each revealed thousands of galaxies sparkling like gems in an infinite jewel box. Some were relatively close to our Milky Way, while others were near the edge of the visible universe. All of the galaxies were seen within patches of the sky the size of a grain of sand held at arm's length. These pictures were humbling and exhilarating, for they were our most detailed glimpses into the depths of the cosmos.

We Are STARDUST

When we see pictures of those distant glowing islands of stars, it's natural to wonder about the origins of matter. Where did the ingredients of galaxies come from—and how did they ultimately give rise to planets and living things? Astrophysics has provided the most successful explanation so far of the creation of the elements, a process called nucleosynthesis.

Two significant phases of nucleosynthesis mark the history of the universe. The first occurred just after the Big Bang, the universe's fiery origin. At first the temperature of the universe was too hot for any matter to exist separate from energy. But as the temperature fell below a trillion degrees Celsius, protons and neutrons began to "freeze out" from the superhot, superdense primordial soup. By the time the temperature dropped to a mere 10 billion degrees, the universe had created all of its protons and neutrons, and electrons were beginning to freeze out as well. The cosmos was just one second old.

Over the next few minutes—three at least, 10 at most—all the neutrons either decayed into pairs of protons and electrons or were captured by other protons to form helium, deuterium (a heavy form of hydrogen), and tiny traces of slightly heavier

A Hubble Space Telescope view down a corridor of space 12 billion light-years long reveals spirals like our own Milky Way dominating the galactic cast of thousands—all within a patch of sky no broader than a grain of sand held at arm's length. Elliptical galaxies with older stars appear as reddish blobs, and peculiar-shaped galaxies may mark the sites of intergalactic collisions. This image is of the sky in the Southern Hemisphere.

elements. The protons, nuclei, and electrons then waited for the universe to chill further, just as we might wait for our soup to cool at dinner. But whereas we wait a few minutes before sipping, the primordial matter waited another 300,000 years or so until the temperature dropped to a frosty 4,000 degrees. Only then was it cool enough for the electrons to join with the protons and the heavier nuclei to form atoms, an event called recombination.

This original Big Bang recipe for the cosmos yielded, by weight, about 75 percent hydrogen and 25 percent helium. There was a soupçon of other elements as well, including lithium, beryllium, and one one-thousandth of a percent deuterium. But the Big Bang produced not a single atom of carbon, oxygen, calcium, iron, or any other heavy element. Those ingredients for planets and life were generated by the second and ongoing phase of nucleosynthesis in the universe. The factories that churn out the heavy elements, astrophysicists realized during the twentieth century, are the stars in the sky.

The original set of cosmic ingredients spawned the first stars. Gravity pulled together pockets of matter that started with slightly higher densities, a slow process that probably took several million years. In time these prenatal stars became hot and dense enough at their cores to ignite the fires of thermonuclear fusion. Hydrogen atoms in the cores fused into new atoms of helium, losing a tiny amount of mass along the way. That "lost" mass became energy in accordance with Einstein's famous equation, $E = mc^2$. The energy flowed out from the cores of the newborn stars, providing the heat and pressure needed to balance any further contractions from the inward force of gravity. This continual balancing act of gravity versus nuclear fusion characterizes the lives of adult stars.

All stars are large balls of electrically charged gas, called plasma. Perhaps because of science fiction–style plasma ray guns, most people think of plasmas as exotic, energetic, and hot. That's true in a star, but in fact plasmas also are common on Earth. When we touch a metal doorknob after shuffling our feet across a carpet, that small spark we feel is an electrical charge running through a plasma pathway in the air that arose milliseconds before the shock. The same process occurs with lightning, albeit on

Nearly 10,000 stars fill this portion of the Large Magellanic Cloud, a nearby galaxy gradually spiraling into the Milky Way. Despite the crowded appearance of the field, the stars themselves are so far apart that smash-ups between the two galaxies' stellar inhabitants are highly unlikely.

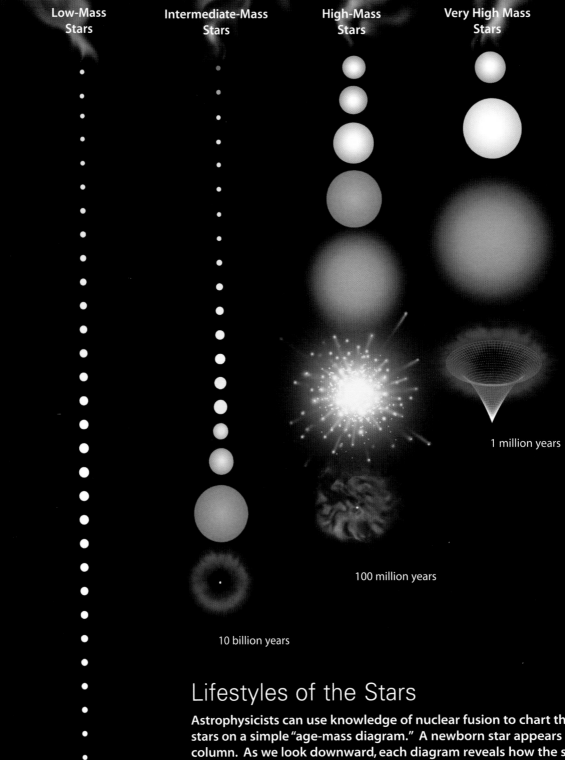

Very Low Mass Stars

Low-Mass Stars

Intermediate-Mass Stars

High-Mass Stars

Very High Mass Stars

1 million years

100 million years

10 billion years

1 trillion years

100 trillion years

Lifestyles of the Stars

Astrophysicists can use knowledge of nuclear fusion to chart the life histories of stars on a simple "age-mass diagram." A newborn star appears at the top of each column. As we look downward, each diagram reveals how the star changes with age until, at the bottom, it dies. The column on the left depicts very low mass stars, which start life with less than one-tenth of the Sun's mass. These midgets never achieve ongoing nuclear fusion in their cores, so they change very little as time passes. They spend trillions of years as failed stars known as brown dwarfs. Low-mass stars, which start their lives with somewhat less mass than that of the Sun, do achieve fusion but they ration their energy supply for hundreds of billions of years before losing their outer layers and dying as white dwarfs. The middle column features intermediate-mass stars such as our Sun. These stars, which range up to 10 times the Sun's mass, pace themselves with only slightly less economy than their low-mass brethren. They fuse hydrogen in their cores for billions of years before swelling to become red giants and then laying bare their white dwarf cores. The final two columns display the most profligate star types: high-mass stars, which range from 10 to 20 times the mass of the Sun, and very high mass stars, from 20 to a gargantuan 100 times the Sun's mass. These giants burn their fuel at a furious clip. In less than 100 million years, they undergo catastrophic gravitational collapse. They then explode as supernovas, leaving behind neutron stars or black holes, or they collapse directly into black holes.

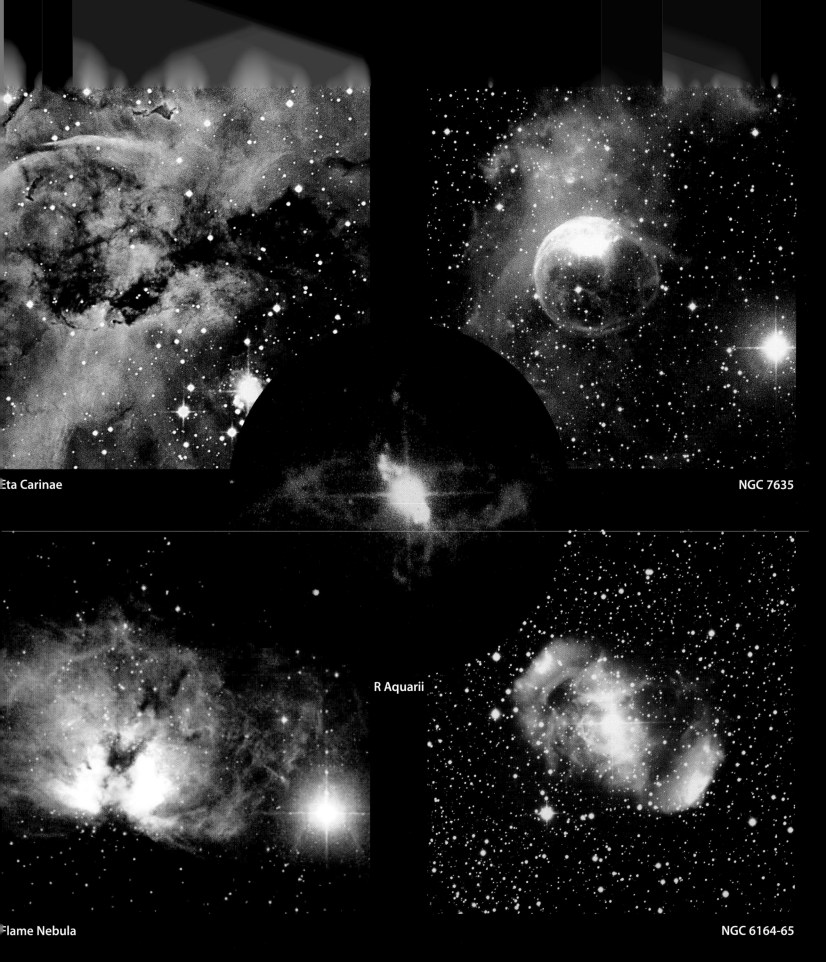

Eta Carinae

NGC 7635

R Aquarii

Flame Nebula

NGC 6164-65

A Stellar Menagerie

From the swirling gaseous loops of NGC 6164-65, the product of gravitational interactions among the trio of stars that make up the system, to the organic looking complexity of the Carina Nebula, the skies abound in stars living out and ending their varied lives. At the center of R Aquarii, for example, a white dwarf draws in material from a companion red giant, occasionally ejecting some of it as strange loops.

a much larger scale. The shimmering glows of auroras in the sky near the North and South poles are plasmas as well, created when the Sun's solar wind—itself a plasma—interacts with charged particles in our upper atmosphere.

Plasmas in space react to changes in temperature and pressure in the same way that a balloon full of gas does on Earth. If you leave a balloon in the hot sun, its internal pressure increases as the gas molecules inside move faster. As a result, the balloon expands. In a refrigerator or freezer, the balloon shrivels as molecules slow down and gas pressure decreases. Similarly, the internal pressure of a star must maintain a constant balance. If it doesn't, the star will either collapse or expand, but in a much more dramatic fashion than our balloon. The critical factor is the amount of outward pressure produced by fusion of the nuclear fuel in the star's core. If the energy production within a star changes even slightly, dramatic evolution occurs.

Most stars fuse their hydrogen happily for billions of years before they run out of fuel. Our own Sun has a comfortable life expectancy of about 10 billion years. But the most massive stars live fast and die young. Their inward gravitational forces are so strong that nuclear fusion must proceed at a breakneck pace to support the stars' weights. Their stores of hydrogen dwindle in just a few tens of millions of years, a cosmic wink of an eye. Contractions in the inner layers then increase the internal pressures enough for other elements to fuse into heavier offspring: Helium creates carbon, carbon begets oxygen and neon, oxygen forges silicon, and silicon sparks iron. Iron is the end of the line, since its fusion absorbs rather than releases energy. The energy output from the star's core then drops precipitously, and the outer layers rush inward. Within seconds the core collapses and unleashes a stupendous shock wave that explodes the rest of the star.

These blasts are called supernovas. They happen about once a century in our galaxy and once a second in the universe as a whole. A typical supernova spews 100 times more energy into the cosmos in its first 10 seconds than the Sun will emit in its entire lifetime. This frenzied burst of energy triggers an orgy of alchemy. The doomed star's elements combine with a blizzard of free protons and neutrons to create cobalt, copper, gold, uranium, and the other heavy elements. The next time you glance at your wedding band or swallow that zinc supplement, offer thanks to some star that exploded in our galactic neighborhood long ago.

As generation after generation of massive stars blow up and scatter their atoms into space, the percentage of heavy elements in the gas clouds of a galaxy increases. In other words, stars relentlessly convert the universe's original supply of hydrogen and helium into the rich panoply of elements that fill out the periodic table. But don't worry; the cosmos won't run out of stellar fuel anytime soon. Even after 13 billion years, stars have burned much less than 1 percent of the primordial hydrogen and helium. New stars will continue to form for hundreds of billions of years. However, the rate of starbirth will gradually decrease as the universe grows older and matter spreads out.

Astronomers can chart subtle changes in stellar ingredients to learn how our Milky Way galaxy has evolved. Newborn stars have the highest proportion of heavy elements because they arise from matter already burned and expelled by many other stars. On the other hand, old stars that formed early in the galaxy's history consist of hydrogen, helium, and very little else. We can detect these elemental signatures as faint patterns imprinted upon the rainbows of light from stars when we examine them with spectrographs.

Surveys of the Milky Way show that the youngest stars reside in the galaxy's gas-rich spiral arms. There, density waves—the ones akin to stop-and-go traffic on our highways—trigger ongoing bursts of starbirth. The Orion Nebula and its quartet of huge baby stars, called the Trapezium, inhabit a spiral arm of the Milky Way next to our own. The oldest stars live in globular clusters, swarms of perhaps 100,000 suns that surround the center of the galaxy like moths around a street lamp. These stars are among the most primitive objects known, with ages rivaling that of the universe. We have yet to spot any stars that contain no heavy elements at all. If we see such objects, we will have found survivors from the first generation of stars after the Big Bang itself.

The creation of heavy elements inside stars has another important consequence for the cosmos: It allows rocky planets to form (page 95). Planetary systems appear to be a natural byproduct of starbirth. Smaller clumps of gas and dust collapse into planets within the thick disks that surround baby stars. However, planets around the earliest stars could draw only from hydrogen and helium. Those first planetary systems thus may have contained bodies like Jupiter but no Earths. Only after generations of massive stars had exploded in nearby parts of the galaxy could rocky planets arise. Models of solar-system formation suggest that dense Earth-like planets probably form in the warm environment close to the parent star. Gas giants, whose elements are more prone to being boiled off by a hot young star, prefer a cooler setting farther away. Recent discoveries of Jupiter-sized planets orbiting close to other stars may conflict with that model. Alternatively, it could mean that over time giant planets can migrate inward toward their parent stars.

When a planet coalesces from the debris around a star, it starts as a thorough mixture of the matter in that part of the disk. But the densest matter quickly sinks to the center of the new planet, a process similar to pulp settling to the bottom of a glass of freshly squeezed orange juice. In Earth's case those materials were iron and nickel. This segregation of dense elements released a tremendous amount of heat, keeping Earth molten for much of its early history. The less dense elements, such as oxygen, stayed near the surface. They eventually formed the floating slabs of rocky crust on which we live. As evidence of that process, Earth's solid surface is nearly half oxygen by weight.

Another source of heat within Earth remains important today: radioactivity. For quantum mechanical reasons, most combinations and sums of protons and neutrons in an atomic nucleus are unstable. Atoms containing such combinations will eject a helium nucleus of two protons and two neutrons (an "alpha particle") or change a neutron into a proton and an electron (a "beta particle"). These decays are mediated by the weak nuclear force. Over time they transmute a radioactive substance into other elements, atom by atom. For example, radioactive uranium decays to lead, a stable element, via a chain of other radioactive elements, including thorium, radium, and radon. If dangerous levels of radon gas build up in your basement, you know that this process is happening in the soil and rocks around your home. There's also a clear benefit from radioactivity: Without it, Earth's interior would be much colder. The planet would be less active and, perhaps, less hospitable to life.

We cannot predict when a particular radioactive atom will decay. Rather, we use statistical averages to calculate overall rates of decay for large numbers of atoms. The time it takes half of any given amount of an element to decay is called its half-life. After two half-lives have passed, one-quarter of the original element is left. After 10 half-lives, only a thousandth remains. These intervals of time can range from fractions of a second to billions of years, depending on the element.

Alchemy by Supernova

From the study of hundreds of supernovas in distant galaxies, scientists have identified two main varieties of these cosmic cataclysms. Type I supernovas are thought to involve gravitationally bound pairs of stars. The slightly more common Type II explosion typically involves a single high mass star—anywhere from eight to 20 times the mass of our Sun. Fusing hydrogen atoms at the rate of some 20 trillion tons per second, such a star reaches the end of its supply in only about 10 million years. As long as fusion lasts at the core, its outward pressure counteracts the inward gravitational pull exerted by the star's enormous mass. As soon as fusion ends, however, the countdown to the explosion begins. Illustrated here and on the following page is the process known as core collapse, which frees enough nuclear energy to blow up the star's outer layers. The process also creates all the elements heavier than hydrogen and helium, seeding the universe with matter to make up not only the next generation of stars, but also planets, moons, comets, and all living things.

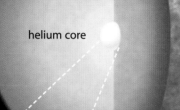

hydrogen and helium envelope

helium core

At the Star's Core

With no more hydrogen left to fuse into helium, nuclear reactions halt, gravity squeezes atoms closer together, and the core heats up. Radiating heat expands the outer envelope to 100 times its original diameter.

helium

carbon/oxygen

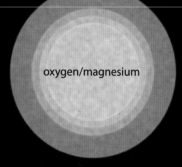

neon/ magnesium

oxygen/magnesium

One Million Years to Blast Off

Gravitational collapse in the core pushes temperatures over 170 million degrees. Helium atoms fuse to form carbon and oxygen.

A Thousand Years to Go

Once helium in the inner shell is used up, the core begins to contract again, in alternating cycles of fusion and contraction. Carbon fuses to neon and magnesium.

Seven Years Left

As temperatures reach 1.5 billion degrees in the stellar core, neon atoms fuse to form more oxygen and magnesium.

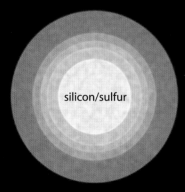

silicon/sulfur

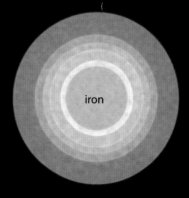

iron

maximum scrunch

A Year to Go

Temperatures in the collapsing core top 2 billion degrees. Compressed oxygen atoms fuse to form silicon and sulfur.

A Few Days Left

At 3 billion degrees, silicon and sulfur fuse to form a compressed ball of iron. This is the last reaction that can take place.

Only Tenths of a Second Left

At nearly 45,000 miles per second, the iron core crushes in on itself, packing an Earth-sized object into a ball 10 miles across. Astrophysicists call this "maximum scrunch."

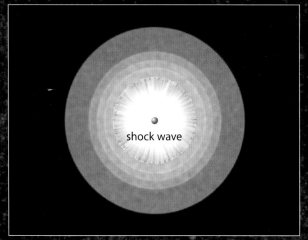

The Explosion Has Begun

Repulsive force between the nuclei overcomes gravity and the iron core rebounds. The explosive shock wave blasts through the layers, creating new heavy elements as it goes.

Seconds into the Explosion

Neutrinos trapped in the core stream out through the star's layers, driving the shock wave to blast off the star's outer layers. The neutrino burst is the first detectable sign of the star's demise.

Hours Later

The shock wave bursts through the star's outer surface, hurling into space several times the Sun's mass in heavy elements. The explosion leaves behind a neutron star, the original star's crushed core.

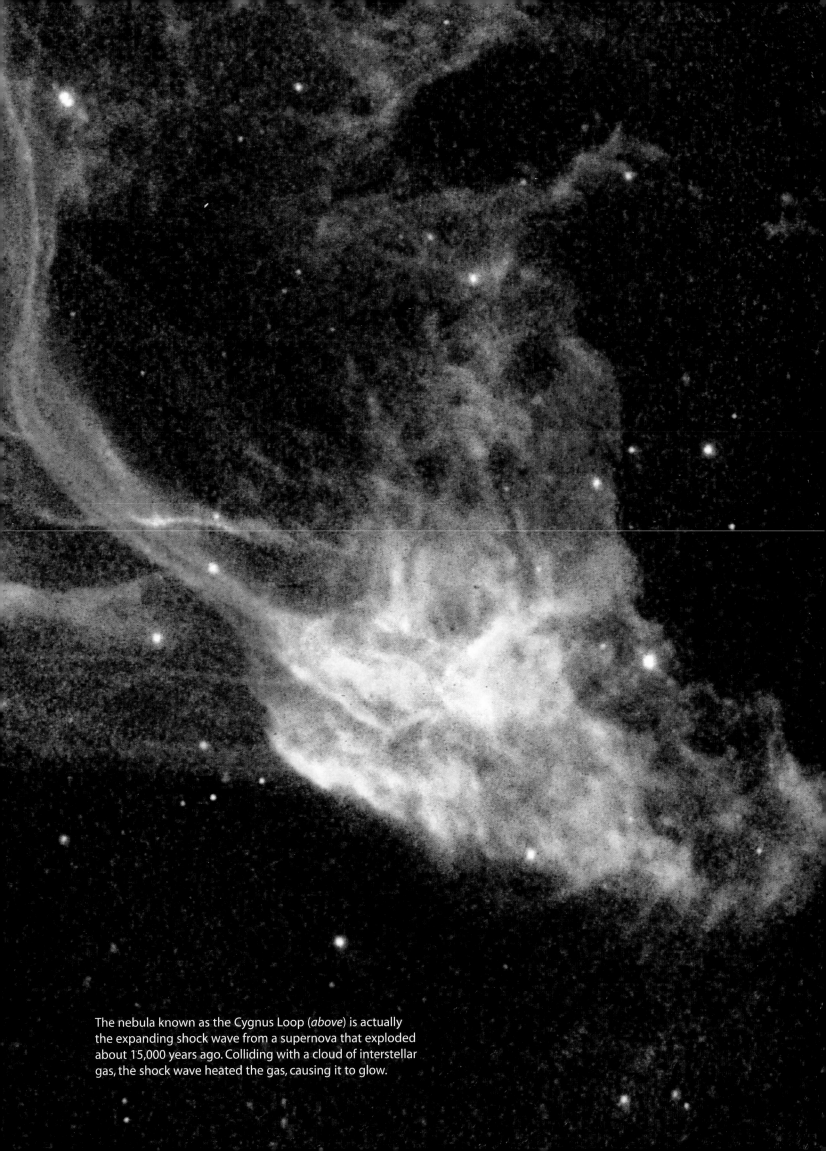

The nebula known as the Cygnus Loop (*above*) is actually the expanding shock wave from a supernova that exploded about 15,000 years ago. Colliding with a cloud of interstellar gas, the shock wave heated the gas, causing it to glow.

Radioactive dating of rocks from the Moon shows that it formed at about the same time as Earth. But the manner of that formation remains a topic of hot debate.

Radioactive decay offers a powerful tool with which we can gauge the ages of various objects. For example, carbon-14 is a radioactive type of carbon that accumulates in living tissues in a known proportion compared with the stable and much more common carbon-12. When an organism dies, it stops adding carbon to itself. The carbon-14 begins to decay with a half-life of 5,700 years, while the carbon-12 remains exactly as it was. Thus, the ratio of carbon-14 to carbon-12 in an old object reveals how long ago it died. Archeologists and paleontologists use this technique, called radiocarbon dating, to estimate the ages of human and other animal remains. They also apply radiocarbon dating to long-dead plants and trees, as well as cloth, paper, and charcoal found at historical digs.

Geologists and planetary scientists use a similar approach to measure the ages of rocks and meteorites. But their elements have considerably longer lifetimes. Potassium-40 has a half-life of 1.26 billion years, while that of uranium-238 is 4.46 billion years. Studies of those elements have shown that meteorites—remnants of the early solar system—are about 4.6 billion years old. The narrow range of ages calculated for these objects suggests that they all formed within an interval no longer than 100 million years.

Radioactive dating of rocks from the Moon shows that it formed at about the same time as Earth. But the manner of that formation remains a topic of hot debate. Some astronomers propose that the Moon was a large asteroid captured gravitationally by the young Earth. Other theories hold that they coalesced side by side or that Earth spun quickly enough in its youth to cast a large chunk of itself into orbit. However, various clues now point to an even more striking origin (page 96). Analysis of rocks collected by Apollo astronauts suggests that the Moon's composition is similar to that of Earth's mantle, the thick rocky layer beneath the crust. More recent space missions indicate that the Moon also has a small iron core, composing just a small percentage of its mass. The best explanation seems to be that a Mars-sized object smashed into Earth while it was still forming. The impact flung an enormous disk of molten rocky debris from Earth's mantle into space, along with a small amount of iron. The Moon then assembled from the debris, leaving Earth battered but no longer alone in the solar system. Computer simulations show that this violent process is physically plausible and may be relatively common in young planetary systems.

The Physics of DENSE MATTER

The rocks that compose Earth and the Moon are products of stars that exploded long ago in our region of the Milky Way. But when stars die, they do more than simply shed matter that enriches the galaxy. They also leave behind the most bizarre states of matter in the universe: ultracompact objects with intense gravitational fields. These stellar cinders are unimaginably dense, denser than anything we could possibly create on Earth. We study them by observing the exotic radiation they emit and their violent interactions with other matter in the neighborhood.

Three fates are possible for the cores of dying stars. Astrophysicists have coined simple phrases that describe the essence of each type of object: white dwarf, neutron star, and black hole. The members of this heavyweight trio share some attributes. For example, the crushing power of gravity forges them in the hearts of stars that have consumed their nuclear fuel. They all attract matter like powerful drains, often with explosive results. Furthermore, each consists of "degenerate" matter. This has nothing to do with socially objectionable behavior. Rather, degeneracy means that the usual spaces between atoms are squeezed into nothingness.

The key difference among the three objects is the degree to which such squeezing occurs. The strength of the cosmic vise is set by the size of the doomed star: Massive stars create more compact remnants. Imagine crushing your car in a futuristic automobile compactor with three settings. The white dwarf setting might spit out a cube of metal and glass the size of a sugar cube but with the mass of your original car. Point the dial to neutron star and you would get a 1-ton speck the size of an amoeba. If you really hated your car and wanted it compressed to black hole densities, it would vanish from sight entirely, collapsed within a volume smaller than an atomic nucleus.

To achieve these dramatic states, stars must overcome the forces of repulsion that control ordinary matter. We can devise a mental picture of how that happens by shrinking our Superdome model of the atom to a more manageable size. Consider the atoms in everyday solid matter—a brick, for instance—as Ping-Pong balls. On that scale each atom's nucleus would be like a grain of talcum powder in the center of the ball. The ball itself is akin to the atom's shell of electrons. Although not quite rigid, the electron shell carves out a volume of space that other atoms may not enter.

In ordinary solid matter, atoms bind together in gridlike structures called lattices. The space between each atom is large compared to the sizes of the individual atoms. It's as though a network of interconnecting toothpicks holds the Ping-Pong balls in fixed positions relative to one another. The toothpicks represent the electrostatic repulsion among the clouds of electrons around the atoms. They resist compression by any force we could muster on Earth, even if we ground the brick to dust.

However, stars can turn that trick readily. We know that during its lengthy adulthood, a star like our Sun balances the tremendous inward crush of its own gravity with outward pressure released by its fires of thermonuclear fusion. This equilibrium will exist in our Sun for 5 to 7 billion years more. Observers in our future solar system will see the Sun's energy output decline temporarily as helium ash builds up in the interior. Then the full weight of the star's outer layers starts to bear down upon the core. For a time the Sun staves off implosion by fusing helium into carbon. This transition pumps out energy at a thousand times the previous rate, resulting in a spectacular swelling of the Sun's outer layers (like our balloon in the hot sunlight). Our home star expands to hundreds of times its current diameter, becoming a "red giant" that engulfs Mercury, Venus, and possibly Earth and Mars within its scorching atmosphere.

Sadly for the Sun, those helium fires die down in about 100 million years. Then, gravity wins in a dramatic fashion. The core collapses under the pressure, as surely as a steamroller flattens an egg. In our analogy the Ping-Pong ball atoms pour into the center of the star. The force of gravity shatters the toothpicks between the atoms until they push against one another in a dense globe—far more densely packed than any state of matter on Earth.

At that point the basic rules of quantum mechanics prevent a further collapse. Electrons in the star's core obey the Pauli exclusion principle: As the gravitational bear hug crams the electron shells closer together, the electrons orbit their atoms ever more excitedly to avoid falling into the same physical state. This creates a new kind of outward pressure. The shells of the Ping-Pong balls grow vastly more rigid than occurs in normal matter. Finally, when the star shrinks to an object the size of Earth,

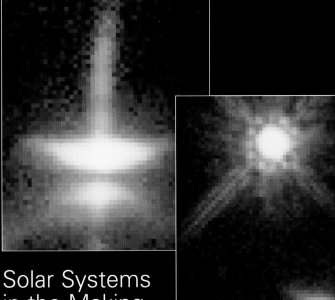

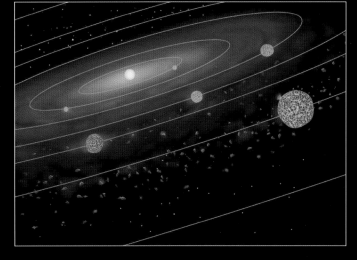

Solar Systems in the Making

The formation of planets around stars is believed to start with the gravitational accretion of a disk of material. Current theory for the formation of our own solar system holds that a large gas cloud collapsed to form the Sun and that the planets accreted in time from a disk of leftover material swirling around the new star.

Astronomers have identified many young stars, such as the two shown here, that are likely candidates for forming planetary systems. Each is surrounded by a disk of orbiting dust and gas, the material remaining from the birth of the star itself. The disk of Herbig-Haro 30 (*above left*) spans 40 billion miles and emits powerful gaseous jets. HK Tauri (*above right*) is actually a binary star system. The dark disk surrounding one member of the pair is 20 billion miles in diameter.

Although these systems may be too young for disk material to begin accreting into planetary bodies, scientists have found evidence for a number of other extrasolar planets in recent years (page 140). So far, none appears to resemble the small rocky planet on which we live, but depicted at right is one scenario for how rocky Earth-like planets might form from the debris around a newborn star.

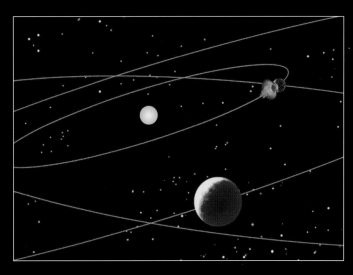

Dust in a protoplanetary disk accretes into rocks that collide and merge into ever-larger bodies (*top*). After about 100,000 years, some of these embryo planets may have up to 70 percent the mass of Earth (*middle*). As larger bodies pull smaller ones into elliptical orbits, violent impacts and mergers occur (*bottom*), venting volatile gases that may eventually form the new world's atmosphere.

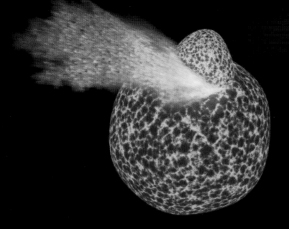

A Mars-sized impactor smacks the embryo Earth, heating and deforming both bodies and spewing ejecta into space.

The Moon's Violent Birth

About 4.5 billion years ago, somewhere between the present orbits of Earth and Mars, a planet about the size of Mars probably struck the embryonic Earth in a collision that ultimately created Earth's only satellite. Known as the giant impact theory, this scenario largely accounts for the particular chemical composition of both bodies. It explains, for instance, why the Moon has only a tiny metallic core and Earth has a considerable one. It also accounts for the amount of angular momentum in the Earth–Moon system. (Angular momentum is the measure of motion of objects in curved paths. In this case it means the spin of each body plus the orbital motion of the Moon around Earth.) As shown here, the young Earth was probably almost completely molten during this process.

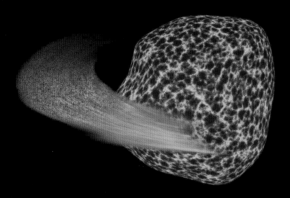

The impactor rebounds and hits Earth again. Most of its metallic core gets incorporated into Earth's core.

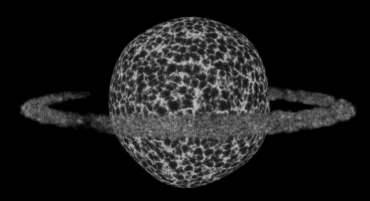

An orbiting ring of very hot ejecta, very little of it metallic, eventually cools and condenses into discrete particles.

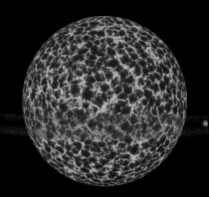

As particles accrete, they sweep up the disk of ejecta. Within about 10 years, the largest body sweeps up the remaining debris to become our Moon.

this pressure halts gravity's march. The dwarf star glows with a fierce white-hot light from the energy of gravitational collapse and leftover thermonuclear fusion.

Astronomers discovered white dwarfs in 1862 when they spotted a dim star orbiting Sirius, the Dog Star. Called Sirius B or the "Pup," the white dwarf is 10,000 times fainter than Sirius, the brightest star in the sky. By applying Kepler's laws of planetary motion to the two stars, we can deduce that Sirius B contains almost as much mass as the Sun and is nearly 5,000 times denser than lead. It will glow for billions of years as it radiates its heat into empty space. White dwarfs behave exactly like coals in your fireplace: They turn orange, then red, and then black as they inexorably cool. We know of thousands of them, but billions surely populate the galaxy. Since all they do is cool down, they are the closest things to cosmic chronometers that we have for estimating the age of the Milky Way.

A white dwarf can flare back to life explosively if it orbits around another star. Sirius and Sirius B form such a binary pair, but the stars are too far apart to interact directly. However, many other pairs of stars in the galaxy orbit each other more closely than the Dog Star and its Pup. If one star becomes a white dwarf and the other is a bloated giant, the dwarf can pull some of its companion's atmosphere onto its surface. This material—mostly hydrogen—can coat the dwarf in a layer thick enough to ignite in a thermonuclear flash. Such stellar hydrogen bombs are called novas. They can flare up every few weeks to every few centuries as new layers of hydrogen accumulate. In spectacular cases the thermonuclear flash can spark the entire white dwarf to collapse and then explode in a supernova that shines as brightly as its entire galaxy of stars.

A star much larger than our Sun suffers a similar fate when it runs out of fuel. If the core of the star exceeds a critical threshhold of mass—1.4 times the mass of our Sun—it spawns a supernova blast. At the heart of the explosion, the Ping-Pong ball shells of the atoms all collapse. Atomic nuclei disintegrate at the centers of the crushed shells. Electrons, neutrons, and protons all jam together, and gravity eradicates the spaces between them. The electrically charged electrons and protons cancel each other out, forming an extraordinarily compact glob of neutrons. A "neutron star" is born. But there's a limit to this squeeze: The pressure of the quarks within the neutrons, a consequence of the strong nuclear force, holds the line. This is neutron degeneracy, matter's last stand against gravity.

A neutron star is as dense as an atomic nucleus. Indeed, we can consider it a gigantic nucleus in its own right. It's downright puny on astronomical scales, measuring only about 10 miles across. Although that's just the size of a large city on Earth, the star contains more mass than our Sun. One teaspoonful of this matter weighs more than 3 billion tons. That's like stuffing a herd of 50 million elephants into a thimble. If we dropped a small piece of neutron star onto the ground, it would slice through Earth like a bullet through cotton and come out the other side.

Most neutron stars enter the universe like whirling dervishes, thanks to nature's laws of motion. Even if the large parent star spun slowly at the end of its life, the conservation of angular momentum dictates that the tiny neutron star must spin more than a million times faster. Such stars usually have intense magnetic fields that drive particles outward along two narrow jets. If the jets happen to point in Earth's direction, we see the neutron star emit blips of radiation each time it rotates. Those flashes, like the pulses from a rotating lantern in a lighthouse, prompted astronomers to call the extreme stars pulsars.

The British radio astronomers Jocelyn Bell and Antony Hewish discovered the first pulsar by accident in 1967. Bell spotted regular fluctuations in the radio signal from an unknown celestial object. Upon further analysis, the astronomers were stunned to find that the source blipped once every 1.33731109 seconds, the steadiest pulsations ever seen. Their first code name for the object was LGM-1, for "Little Green Men," since they half-jokingly proposed that it might emanate from an extraterrestrial civilization. But other pulsars turned up within a few months, including one at the center of the Crab Nebula. That object, which flashes 30 times per second in the middle of an expanding cloud of debris from a supernova, verified Bell and Hewish's suspicion that the pulsars were spinning neutron stars. Since then astronomers have found more than 1,000 other pulsars, almost all of them in our Milky Way galaxy. Most whirl between one and 100 times each second. The fastest yet seen spins an incredible 642 times per second. At that pace a point on its equator moves at one-seventh the speed of light.

The motions of pulsars are so steady that we can time them to within one part in a million billion. That's 15 decimal places, a rotational accuracy that rivals the best atomic clocks on Earth. Occasionally, we see tiny glitches in their rotation speeds—sudden increases that may make the pulsar spin a billionth of a second faster than before. These flaws probably arise when "starquakes" fracture the pulsar's brittle crust, briefly changing its moment of inertia. But unlike earthquakes on our planet, which crack Earth's crust a few yards at a time, a pulsar quake might break its surface by less than one one-thousandth of an inch. That tiny movement creates enough of a change in a pulsar's spin that we can detect it from thousands of light-years away.

A pulsar comes as close as any object in the universe to a perpetual motion machine. However, its spin must wind down as surely as that of any gyroscope or top on Earth. The slowdown stems from a pulsar's intense magnetic field, which exerts a gradual but persistent drag on the spinning object by interacting with nearby material in space. After tens of millions of years of this steady braking, most pulsars spin just once every few seconds. We can no longer see those old pulsars because they spin too slowly to create beams of radiation. As time goes on, they spin more slowly still, although it may take many billions of years for them to stop. Dead and invisible to us, these barely spinning pulsars are the most nearly perfect spheres in the cosmos. No rotational stresses make them bulge, and their fierce gravitational fields flatten any bumps on the surface more than a few atoms high.

The realm of pulsars seems exotic indeed. But for the ultimate in compact-matter weirdness, black holes win the prize. If the core of the collapsing star is more massive than about three times the mass of our Sun, not even neutron degeneracy can sustain its structure. Quarks, the foundation of nuclear matter, crush down upon themselves. In gravity's final triumph, what's left of the star collapses without limit as the fabric of space and time folds in on itself. Only mass, rotation, and electric charge remain where subatomic particles once existed.

Unlike other kinds of objects, black holes do not merely wrinkle the fabric of space-time. They rip it, permanently. Everything going into a black hole falls through a hole in the space-time cloth. According to Einstein's general theory of relativity, we can represent space as a rubbery sheet. Objects make dimples in the sheet and stretch space-time toward their centers. A black hole is aptly named, for it behaves like an infinitely deep dimple. If something falls into it—including a beam of light—it won't come back.

The edge of the black hole, the point of no return, is called the "event horizon." A black hole's mass is the primary factor that determines the size of its event horizon.

If we dropped a small piece of neutron star onto the ground, it would slice through Earth like a bullet through cotton and come out the other side.

(A rapidly spinning black hole has a smaller event horizon than a stationary black hole of the some mass.) That dimension is called the Schwarzschild radius, for the German astrophysicist Karl Schwarzschild. For example, a collapsing stellar core with three times the mass of our Sun would form a black hole with a Schwarzschild radius of about 5 miles. A black hole with the mass of Earth would have a Schwarzschild radius of less than half an inch. Its density would be 1.5 trillion times greater than that of a neutron star. A bathtub full of these gumball-sized black holes would outweigh all of the matter in our solar system.

One does not need to cross the event horizon of a black hole for bizarre things to start happening. Imagine a wayward astronaut with the misfortune of drifting head-first toward a three-solar-mass black hole. By itself the hole's gravity is no deadlier to the astronaut than if he were diving into a swimming pool on Earth. After all, one is always weightless in free fall. The black hole's tides are a different story. As our wanderer falls farther toward the center of the hole's gravity well, he feels a vastly stronger pull on his head than on his feet. Within 10 miles or so of the black hole, the tides tear his body to pieces. The forces extrude him through space like toothpaste being squeezed from a tube. When he reaches the event horizon, his former body is no more than an extended string of atoms.

The best evidence for black holes with a few times the mass of the Sun comes from binary star systems, in which one star orbits an invisible but hungry companion. Gas drawn off from the visible star spirals into the companion. This disk of material accelerates to fantastic velocities as it slides down the black hole's space-time vortex. Friction within the disk heats up the gas to hundreds of thousands of degrees. The gas glows blue-hot and emits floods of ultraviolet radiation and x-rays. Although the black hole itself eludes our view, we can deduce its presence because a gaseous speedway encircles it, ablaze with high-energy radiation. Astronomers also have found strong signs that supermassive black holes—monsters with millions or even billions of times the mass of our Sun—lurk at the cores of most galaxies. These objects probably produce some of the most energetic outbursts in the cosmos (page 146).

On the other end of the size spectrum, some cosmologists believe the universe may contain swarms of "mini" black holes. Conditions might have been just right during the first moments of the Big Bang to compress small clumps of matter into

A Pulsing Heart

At the core of the billowing gas and dust of the Crab Nebula (*right*) is a remarkably metronomic powerhouse known as the Crab Pulsar. First detected as radio sources in 1967, pulsars have proven to be a varied lot, radiating at almost every wavelength of light, including very-high-energy gamma rays. As explained on the next page, these complex objects are believed to be neutron stars, the collapsed and tightly compressed remains of massive suns that have exploded as supernovas. Spinning at the rate of 30 times per second, the Crab Pulsar is what's left of a star whose blazing death throes were witnessed nearly a thousand years ago.

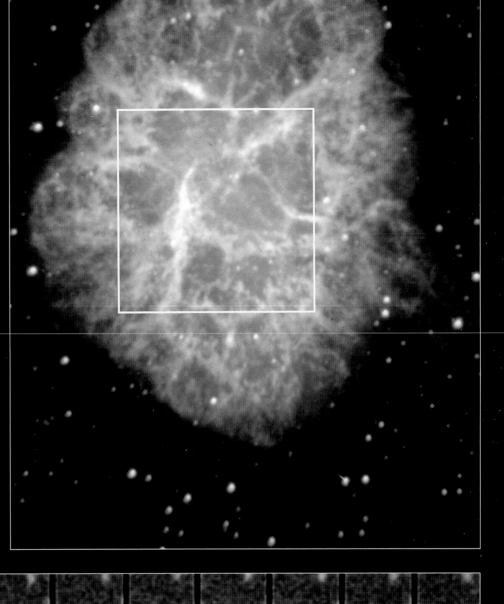

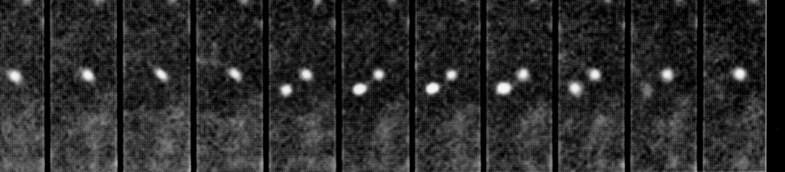

in a supernova explosion. What's left is an object with a mass of about 1.4 times that of our Sun compressed into a sphere about 10 miles in diameter. Because of its vastly reduced size, the neutron star spins more than a million times faster than its parent star, creating a strong magnetic field and emitting energy along its magnetic axis. Observers on Earth see this emitted energy as a pulse because the star's magnetic axis does not coincide with its rotational axis (*right*). Rather, the beams rotate as the pulsar spins, so we only see it if it crosses our line of sight. Sometimes weaker pulses alternate with stronger ones (*graph at right*), indicating that the pulsar's two beams may be of different intensities.

Ticking Every Millisecond

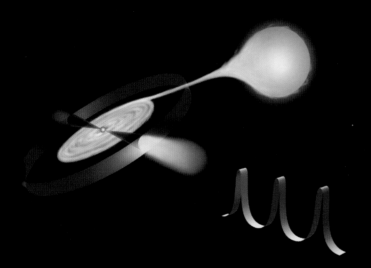
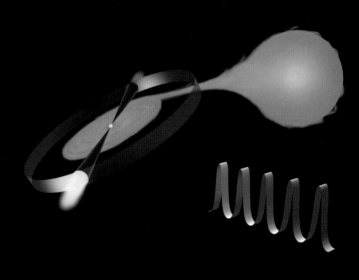

Of the more than 1,000 radio pulsars detected so far, several dozen have been found that spin hundreds of times per second. Astronomers theorize that in order for the neutron star to be "spun up" to such incredible speeds, it must be part of a binary pair. Depicted here is one theory for how a millisecond pulsar comes to be. Once the first star has exploded as a supernova, the resulting neutron star begins to draw matter from its partner (*above*). The companion becomes a red giant (*above right*) as the pulsar continues to draw

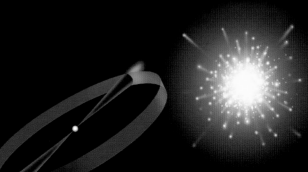

black-hole densities. For instance, a clump with as much mass as an asteroid would form a pinhead-sized black hole. This remains in the realm of theory; we have seen no evidence for such objects. But if they exist, mini black holes could zip through Earth upon occasion. The resulting shock wave in the atmosphere might produce an effect similar to the impact of a small asteroid. A few scientists speculated that the 1908 tree-flattening blast at Tunguska, Siberia, was just such an event. However, further studies suggested that the guilty party was a tiny comet or a rocky asteroid that vaporized in the atmosphere.

Mini black holes would behave in other odd ways. If one wandered through the Sun, it could become trapped by the Sun's gravity and oscillate back and forth through the star for millions of years. Eventually, calculations show, it would stop at the core of the Sun. The Sun would barely care, since the hole's gravitational field would be intense only within a radius of a few feet. Even more strangely, the English physicist Stephen Hawking has suggested that a black hole should evaporate over time via a quantum mechanical process now known as "Hawking radiation." If so, the black hole would emit x-rays and gamma rays for billions of years at an ever-increasing rate. Ultimately, it would explode in a burst of pure energy. It's hard to imagine a more complete transformation: a chunk of the densest matter of all vanishing with a flash into the void.

Too Much Matter

A neutron star is one possible end point of the demise of a massive star. If the star's core is very massive—more than three times the mass of our Sun—even neutrons can't survive the crushing power of gravity. As the core shrinks, its gravitational pull becomes unstoppable. The core collapses in on itself and vanishes as the rest of the star explodes in a titanic supernova. The core becomes a black hole that emits no visible light. Indeed, all of the familiar characteristics of stars, including color, luminosity, and chemical composition, no longer exist. Black holes retain only three properties: mass, spin, and electric charge.

Incredible as all this sounds, the laws of gravity as we currently understand them predict just this. Theoretical physicists have developed several models that describe different varieties of black holes. One type, known as the Schwarzschild black hole (*near right*), is characterized by its mass alone. All of the mass concentrates at an infinitely dense point known as a singularity, located at the center of the hole's "surface"— the event horizon. However, another type of black hole, the Kerr black hole, is probably much more common in the universe. It is simply a Schwarzschild black hole that spins.

Schematically illustrated on the opposite page, the Kerr black hole's rotation forces the singularity to take the shape of a ring within the event horizon rather than a point. The rotation also creates a donut-shaped region where space warps so severely that the path of a light beam will close back upon itself and trace a stable orbit around the hole. If light approaches closer than this zone, it spirals inward helplessly toward the event horizon. Such is the inescapable fate of light and anything else that wanders too close to the black hole's deep gravitational well.

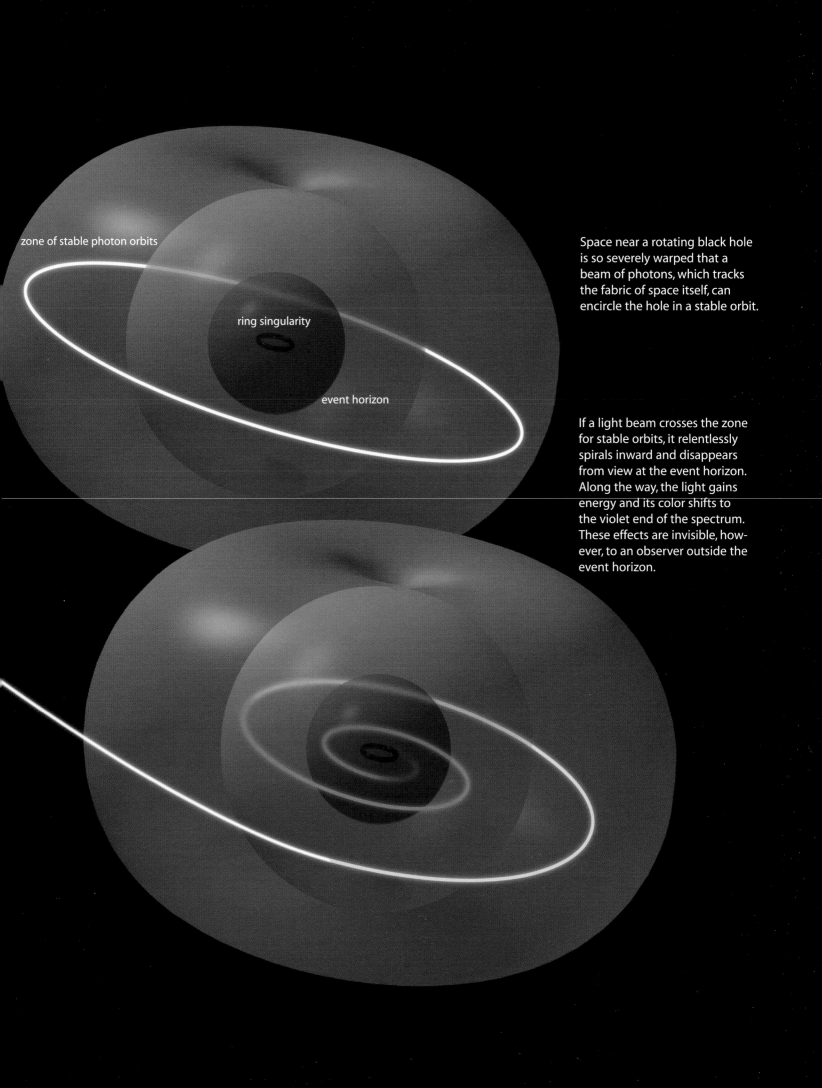

zone of stable photon orbits

ring singularity

event horizon

Space near a rotating black hole is so severely warped that a beam of photons, which tracks the fabric of space itself, can encircle the hole in a stable orbit.

If a light beam crosses the zone for stable orbits, it relentlessly spirals inward and disappears from view at the event horizon. Along the way, the light gains energy and its color shifts to the violet end of the spectrum. These effects are invisible, however, to an observer outside the event horizon.

A Destructive Dance

Among the more lethal pairings in the cosmos is a binary star system in which one member has become a black hole. One such system, in the area of the constellation Cygnus, is an x-ray source known as Cygnus X-1, discovered in the early 1970s. The visible member of the pair is a blue supergiant with a mass about 30 times greater than the Sun. Stars of this type would not ordinarily emit x-rays, but studies show that hot gas from the star is flowing toward an unseen companion. Other analysis suggests that the x-ray-emitting region is an accretion disk formed when matter from the supergiant is drawn into the rotating black hole.

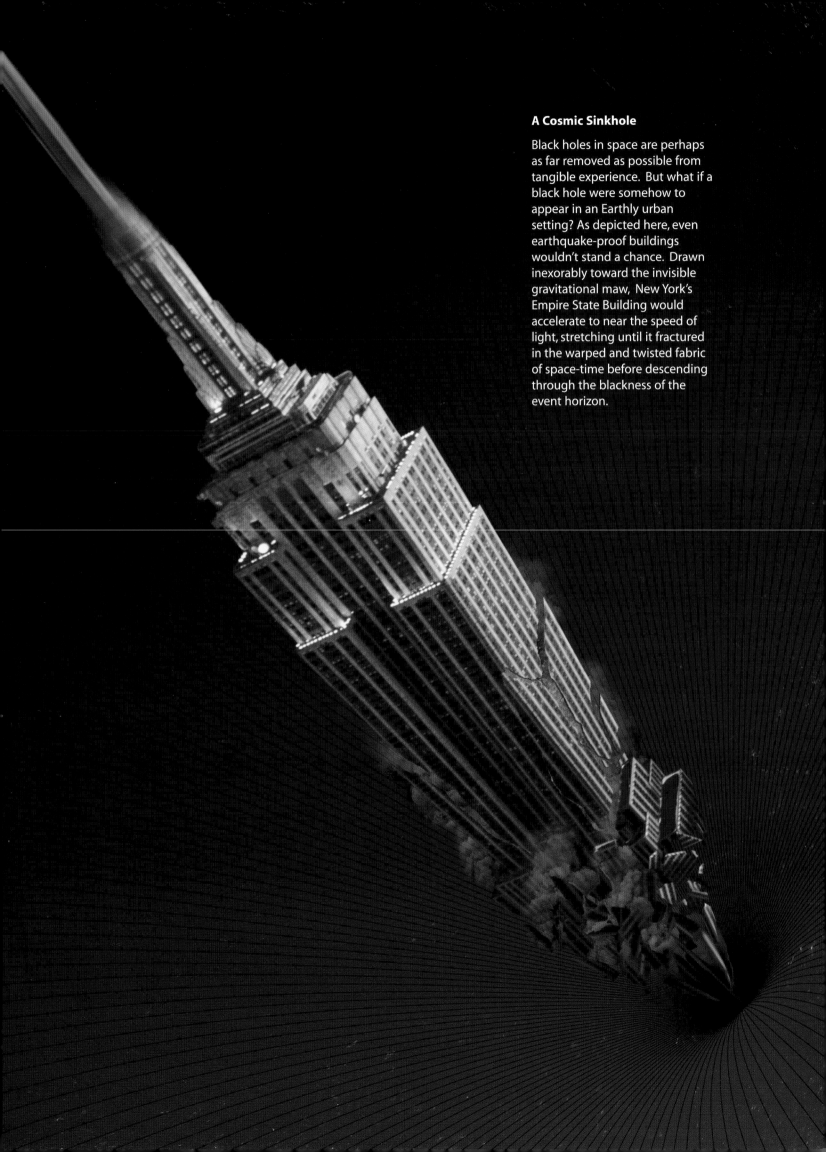

A Cosmic Sinkhole

Black holes in space are perhaps as far removed as possible from tangible experience. But what if a black hole were somehow to appear in an Earthly urban setting? As depicted here, even earthquake-proof buildings wouldn't stand a chance. Drawn inexorably toward the invisible gravitational maw, New York's Empire State Building would accelerate to near the speed of light, stretching until it fractured in the warped and twisted fabric of space-time before descending through the blackness of the event horizon.

Energy
The Power of Cosmic Phenomena

The blazing act of stellar
self-destruction known as
Supernova 1987A released as
much energy as that produced
by more than 200 million Suns.
Shown here 6 months after
it exploded, the supernova
(*below*) gleams as brightly as
an entire cluster of stars at the
heart of the Tarantula Nebula
(*above left*), testament to the
vast stores of energy in the
universe.

Nearly 170,000 years ago a giant star blew up in the Large Magellanic Cloud, the galaxy closest to our Milky Way. Silver, gold, lead, and other heavy elements rifled into space, forged by the fires of the star's spectacular death. Its remnants flew outward at millions of miles per hour to form a hot nebula of glowing gas. Fierce shock waves may have triggered nearby clouds of gas and dust to collapse into baby stars, just as loud noises can set off an avalanche on a snowy mountain slope. The aftermath of the blast rippled through the star's cosmic neighborhood for tens of thousands of years.

Beyond this maelstrom of matter and motion, the star also made a spectacle of itself far and wide with an enormous burst of pure energy. Some of this energy took on a familiar form: visible light, like the light we see from our Sun. This light streamed across the gulf between the Large Magellanic Cloud and the Milky Way, covering nearly 6 trillion miles per year. On February 23, 1987, a tiny fraction of the star's light finally reached Earth. The Canadian astronomer Ian Shelton took a photograph of the Large Magellanic Cloud that night from an observatory in Chile. He noticed a bright spot in the image that hadn't been there the night before. When he went outside, Shelton became the first person in more than a century to simply look up in the sky and see a star blowing itself to bits.

The explosion, named Supernova 1987A, was the brightest supernova seen from Earth since one recorded in 1604 by Johannes Kepler. Outside of the Sun, Moon, and planets, it quickly became one of the most intensely studied astronomical objects in history. Within a few hours it shone as brightly as if it drew its power from 200 million suns. Other forms of light, invisible to our eyes, also taught us about the workings of the supernova. For instance, high-energy x-rays and gamma rays revealed that the inner and outer layers of the star mixed together when it exploded, instead of expanding in smooth shells. Radio waves opened a window into the turbulent heart of the supernova, where a spinning neutron star—a pulsar—may lurk. Despite some tantalizing hints, astronomers have not yet spotted the pulsar's telltale repeating signal.

Supernova 1987A's impressive visual impact landed it on the cover of many magazines. However, most people don't realize that light was just a tiny fraction of the supernova's prodigious output of energy. The doomed star shed 30,000 times more energy in the form of ghostly subatomic particles called neutrinos. In the first

10 seconds of its explosion, the star unleashed 10 billion trillion trillion trillion trillion neutrinos. That's 1 followed by 58 zeros. During those 10 seconds, the detonation produced more power than the combined output of all the stars in the visible universe.

The neutrinos flashed into space in all directions, at or near the speed of light. Because they rarely interact with other matter, the neutrinos zipped through almost everything in their paths—including our planet, once they got here 167,000 years later. Indeed, every square inch of your body was pierced by about 300 billion neutrinos shortly before Ian Shelton spotted Supernova 1987A. Astronomers managed to stop 19 of those fleeting particles in their tracks with two giant underground vats of water, where the neutrinos made tiny flashes. This marked the first detection of neutrinos from an event beyond our solar system. There's an infinitesimal chance that a similar flash occurred within the vitreous humor in one of your eyes at the moment the neutrinos struck Earth.

Supernova 1987A was a nearby example of how energetic our cosmos can get. Supernovas may seem rare, but the universe contains so many stars and galaxies that one pops off somewhere about once every second. In our Milky Way we can expect a star to die in this fashion once or twice a century. A thick pall of dust across the galaxy's center hides many of these stellar bombs. But if one of the stars we see in the night sky exploded, it would outshine Venus at its brightest and might rival the full Moon. Gamma rays from a very close supernova could even strip away the protective ozone layer in Earth's atmosphere and eradicate life on the planet. That would be the ultimate irony, since generations of long-ago supernovas seeded our cosmic neighborhood with the ingredients that eventually gave rise to life on Earth.

ENERGY Powers the Universe

When it comes to energy, you probably don't think of gamma rays and neutrinos from exploding stars. Rather, energy has other meanings during everyday life. When you have no energy, it's an excuse to lie on the couch and watch television. In an energy crisis, it's wise to drive less and turn down the thermostat. A high-energy rock concert makes your ears ring. Our common concepts of energy add seemingly unscientific twists to motion, power, loudness, or some physical activity.

In the world of science, such references are not far off the mark. Energy is defined as the capacity to do work—to move objects from here to there or to change their configuration somehow. Nature features as many varieties of energy as does our informal language. We use mechanical energy to pick things up and put them down. Electrical energy lights our homes and runs our appliances. The chemical energy produced in our cells keeps our bodies going. Moving objects have energy, and so do unmoving ones. Noise is energy; heat and light are energy. Strangely enough, even matter is energy. These interrelationships show why motion, matter, and energy are so closely linked when it comes to describing the universe. The diverse forms of energy also explain why studying it has posed such a challenge to physicists in the past three centuries. The importance of the quest to understand energy in its many guises cannot be overstated. After all, energy powers the universe.

Energy has a place alongside momentum as a fundamental conserved quantity. The physics of the world around us depends on the fact that the total amount of energy within a system remains the same, unless an outside force acts upon it. Energy is never created or destroyed. Any time it seems to disappear, it has merely changed into another form. We can use this critical principle to track the peaks and valleys of

Elusive Stellar Messengers

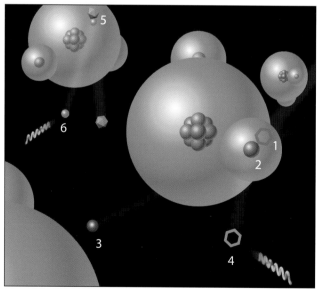

When a blue supergiant star in a neighboring galaxy exploded into history as Supernova 1987A, it spewed countless trillions of ethereal particles called neutrinos into space. Nearly 170,000 years later 19 of them were detected as the swarm passed through Earth.

The detection was indirect, recording not the neutrinos themselves but the byproducts of their interactions with molecules of water (*right*). Several hours before the supernova was noticed in the sky, the underground Kamiokande detector in Japan recorded the arrival of 11 of these elusive specks. Halfway around the world, a detector in an Ohio salt mine counted eight more.

Though seemingly sparse, these results supported the prevailing theory that a supernova explosion releases more than 99 percent of its energy in the form of these nearly massless particles. Today, other detectors, including a mammoth successor to Kamiokande called Super-Kamiokande (*opposite*), continue to track the passage of these interstellar mites in hopes of understanding their elusive nature and determining their mass.

An electron-antineutrino (*1*) bangs into the single proton that is the nucleus of the hydrogen atom (*2*), yielding a neutron (*3*) and a positron (*4*) that emits a flash of light. At top left, an electron neutrino (*5*) nicks an oxygen atom and knocks loose an electron (*6*) with enough force to produce light. The light flashes are picked up by sensors deep beneath Earth's surface (*opposite page*).

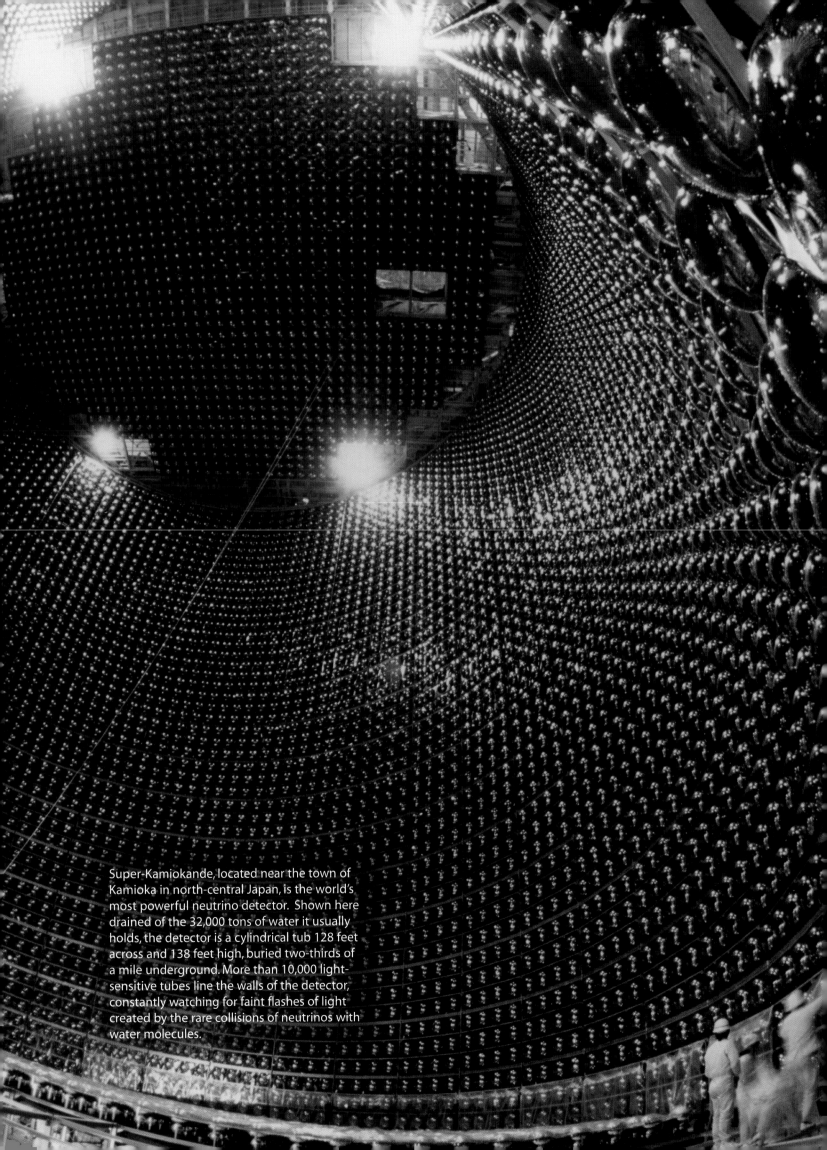

Super-Kamiokande, located near the town of Kamioka in north-central Japan, is the world's most powerful neutrino detector. Shown here drained of the 32,000 tons of water it usually holds, the detector is a cylindrical tub 128 feet across and 138 feet high, buried two-thirds of a mile underground. More than 10,000 light-sensitive tubes line the walls of the detector, constantly watching for faint flashes of light created by the rare collisions of neutrinos with water molecules.

a roller coaster and the ups and downs of temperature in Earth's atmosphere. Because the universe as a whole is a self-contained system, conservation of energy applies to the entire cosmos as well. Energy's shape-shifting abilities make it hard to follow from one form to the next, but a few examples will illustrate how central it is to the workings of our world.

We are most familiar with the energy of motion, called kinetic energy. Everything that moves has kinetic energy. The amount goes up quickly as the speed of an object increases. If you double its speed, it carries four times more kinetic energy. Tripling its speed increases the kinetic energy by a factor of nine, and so on. This simple relationship means that tiny objects can carry a lot of kinetic energy if they travel fast enough. For instance, imagine the difference between trying to catch a pellet thrown to you and one fired from an air gun. In space, a collision with something as small as a fleck of paint can inflict serious damage on a satellite or the windows in the Space Shuttle because objects in orbit can collide at relative speeds of thousands of miles per hour.

Kinetic energy routinely swaps back and forth with a less obvious type of energy, called potential energy. That switch occurs in the presence of any force that can move objects. Consider an archer's bow and arrow. The archer stores potential energy in the bow by doing work on it—drawing the cord back to pull the flexible bow into a taut "C" shape. Releasing the cord transfers most of that potential energy to the arrow. The arrow then flies away with enough kinetic energy to reach and pierce its target. When a tennis player hits a ball, the combined kinetic energy of the ball and the racket bends the racket's strings and crushes the ball into an oblong shape. In an instant the potential energy held within those distorted shapes converts back into kinetic energy and propels the ball in the opposite direction with renewed vigor.

The Mars *Pathfinder* spacecraft used this same type of exchange in a unique way when it landed on Mars in 1997. To absorb the kinetic energy of impact, large balloons inflated around the vessel before it hit the ground. Most of that energy was briefly stored as potential energy within the compressed balloons. Then *Pathfinder* bounced, hit the ground, and bounced again several times. By the time the kinetic-to-potential energy flip-flop dissipated into other forms (such as friction, dents on the ground, and heat within the balloons), the craft came to an undamaged halt.

The most pervasive form of potential energy on Earth and in the universe is gravitational potential energy. Earth's gravity will accelerate any object toward the planet's center when given a chance. To endow something with potential energy, all you need to do is lift it. If you let go, that potential for motion converts into kinetic energy as the object falls. When the floor gets in the way, the kinetic energy transforms into acoustic energy—thud!—and perhaps enough mechanical energy to crack or shatter the object. If you've ever watched a glass plummet from the edge of your dinner table, you know exactly how this works. In a similar way, engineers at amusement parks design conveyor belts to do work on roller coasters and their passengers. From high atop the first peak, the gravitational potential energy of cars and riders converts into a dizzying rush of acceleration, rotation, and high-speed motion.

More often than not, something prevents an object from gaining speed continuously as it falls. Resistance from air or water is a good example. The converted potential energy then must reveal itself in some other way, frequently as heat. In the mid-nineteenth century, the English physicist James Joule explored this phenomenon with a clever device. Falling weights powered the motions of rotating paddles, which in turn stirred a jar of water. The potential energy of the weights could not transfer entirely into kinetic energy to drive the paddles because the water retarded those motions. As a result, the water warmed slightly. The same effect should apply to water plunging 160 feet over Niagara Falls, Joule reasoned. He calculated that resistance to freely falling motion in Niagara Falls should heat the water by one-fifth of a degree. (To recognize Joule's work on heat and the energy of moving objects, we now call one unit of energy a "joule.") Joule could not foresee that his Niagara Falls thought experiment would serve as a perfect analogy for gas heating up as it falls into a black hole. The cosmic distances and temperatures in such a system—billions of miles and millions of degrees—lead to bursts of energy that we can detect across the universe (page 150).

The conversion of energy from potential to kinetic to heat happens all the time, both locally and everywhere in the cosmos. When you drive your car down a steep hill, your potential energy converts to kinetic energy. Apply the brakes and you prevent a dangerous acceleration by converting much of your car's kinetic energy into warming up the brake pads and the rubber tires. The Space Shuttle's controlled reentry into Earth's atmosphere is marked by intense heating of the specially designed tiles that cover it. The tiles grow as hot as 3,000 degrees Fahrenheit. They swiftly radiate this

The conversion of energy from potential to kinetic to heat happens all the time, both locally and everywhere in the cosmos.

heat into space instead of transferring it to the shuttle and its crew. That's the only safe way to dispose of what started as the shuttle's gravitational potential energy.

But even at its initial speed of 17,000 miles per hour, the shuttle's journey through the atmosphere is sluggish compared with that of visitors from deep space. When a comet or asteroid approaches Earth, it falls from many millions of miles away. It converts enough of its potential energy into kinetic energy to strike the atmosphere at speeds of up to 50,000 miles per hour. That's dozens of times faster than a supersonic plane. Past encounters between Earth and interlopers from space have revealed the different ways in which such releases of energy can affect our planet.

On the night of June 30, 1908, a seismograph in Irkutsk, Russia, recorded what resembled a distant earthquake. A thousand miles away, villagers near the Tunguska River in Siberia witnessed a fireball streaking through the sky, a burst of light, a thunderous sound, and an enormous blast. Scientists finally ventured to the remote site in 1927 to find nearly 1,000 square miles of burned and flattened forest. Modern impact analysis suggests that a rocky asteroid or the core of a small comet, about 200 feet across, knifed into the atmosphere at a shallow angle and exploded several miles above the surface. The object's kinetic energy converted almost completely into heat, reducing it to dust. The explosion's power was easily a thousand times greater than that of the bomb dropped on Hiroshima. Had it struck over an urban part of the world, everyone in an area the size of New York City would have died instantly.

An asteroid of similar size struck an uninhabited sandstone plateau 50,000 years ago in what is now northern Arizona. That object, a solid mixture of iron and nickel, was denser than the Tunguska impactor. Much of the asteroid melted as it screamed through the atmosphere. The molten storm spread into a fiery plume over the Four Corners region where Arizona, Colorado, New Mexico, and Utah now meet. The rest of the asteroid melted on impact and gouged a circular scar nearly a mile wide and almost 600 feet deep. The rim of this scar, called Meteor Crater, stands 15 stories above the surrounding land—a raised "lip" similar to the one you can produce by spiking a volleyball into sand. Such impacts were common billions of years ago when the solar system was young and crowded with debris, but they are rare today. Even

Mute testimony to the power of gravitational potential energy converted to kinetic energy and heat, Arizona's Meteor Crater is all that's left of an asteroid that struck Earth some 50,000 years ago. The crater is almost 600 feet deep and 2.4 miles in circumference. More than 2 million spectators could watch from the inner slopes as 20 football games were played simultaneously on the crater floor.

Because we see things move and watch them crash, we easily perceive the many manifestations of kinetic energy in the universe. Potential energy and other common forms of energy, however, often elude our senses.

so, comets or asteroids as wide as a football field still strike land somewhere on Earth every few thousand years, on average, with explosive results.

The deadliest asteroid to hit Earth in recent geological history struck about 65 million years ago off the coast of what is now the Yucatán Peninsula in Mexico. Today, a crater more than 100 miles across is all that remains, much of it under the seafloor in the Gulf of Mexico. The 6-mile-wide asteroid carried 10 million times more kinetic energy than either the Tunguska or Meteor Crater objects. The heat produced by an explosion of that magnitude burned the air itself, as well as most vegetation on the continents. It also threw enough dust into the upper atmosphere to block sunlight for many months. The environmental aftermath, rather than the impact itself, led to extinction of the dinosaurs and most other species on the planet.

Because we see things move and watch them crash, we easily perceive the many manifestations of kinetic energy in the universe. Potential energy and other common forms of energy, however, often elude our senses. One form is crucial to the workings of the universe: nuclear energy. The name derives from the nuclei of atoms, bound together tightly by the strong and weak nuclear forces. During radioactive decay, nuclear bonds in unstable atoms break, releasing a flash of stored potential energy. The atoms eject helium nuclei (alpha particles) and electrons (beta particles), which carry away kinetic energy. Radioactive decay also releases energy in the form of gamma rays—energetic light that heats up its surroundings. Such atomic light warms the interior of Earth and other planets. In space, supernovas forge countless radioactive atoms of nickel and cobalt whose subsequent decay enables the explosions to shine brightly for many months.

Especially energetic breakups occur in processes called nuclear fission and nuclear fusion. Fission, the splitting of heavy atoms such as uranium, releases a million times more energy than the detonation of an equivalent amount of TNT. Under controlled conditions, fission drives nuclear power plants. Uncontrolled fission is the energy source for the atomic bomb. Pound for pound, fusion of hydrogen nuclei into helium is 100 times more energetic still. Our deadliest weapons draw their awesome power from this reaction. So too do the Sun and, for most of their lives, all stars. Indeed, stars are brilliant hydrogen bombs cloaked and contained by hundreds of thousands of miles of gas.

Conduction, Convection, and Radiation

Nature provides three ways to transport heat: conduction, convection, and radiation. Conduction is the transfer of heat from one part of a system to another through the vibration of stationary particles.

Convection is the transport of heat by physically moving particles from hot to cold regions. Radiation does what the other two methods cannot: it carries energy through a vacuum. Vibrations of atoms and molecules are converted into the packets of electromagnetic waves called photons—the light reaching us from the rest of the cosmos.

Conduction

In a pot on a stove burner (*left*), heat is carried by conduction through the burner to the pot by vibrating atoms and molecules. Plastic and wood are poor conductors, so they are often used to make the handles of pots, lids, and cooking utensils. But you use a potholder when grasping the hot handle of cast iron cookware. On a hot, sunny day at the beach (*above*), you need insulation to keep your feet from being burned by contact with the sand.

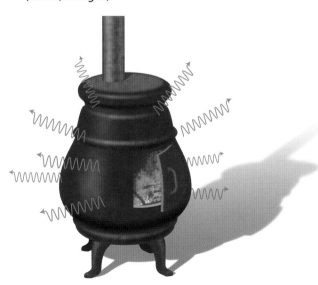

Convection

Convecting fluids move in circular currents called convection cells, as illustrated by bubbles in a pot of water coming to a boil (*above*). Inside the Sun convection cells carry thermonuclear heat to the surface and out into space (*above, right*). High-resolution photographs show the Sun's surface to be granular and bubbly, like a pot of boiling soup (*above, far right*).

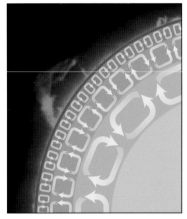

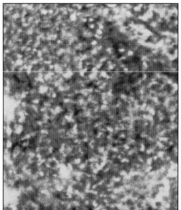

Radiation

Of the entire spectrum of wavelengths found in electromagnetic radiation, the kind we feel as heat lies in a range that readily triggers vibrations and random motions in atomic matter. So heat radiation, commonly known as infrared, is actually a form of light. Just as our skin warms when we sit next to a fire (*left*), interstellar gas warms when heated by radiation from hot young stars. The gas then emits radiation of its own, which we see in so-called emission nebulas (*above*).

Nuclear energies are so enormous because breaking nuclear bonds for heavy elements (such as uranium) and creating them for light elements (such as hydrogen) actually reduces the total mass of all of the nuclei involved. The lost mass converts to energy via Einstein's famed equation, $E = mc^2$. This simple string of characters embodies nature's most mysterious manifestation of energy. It declares that energy is matter, multiplied by a truly gigantic number—the speed of light squared. In other words, a minuscule amount of matter packs an astonishing punch if you convert it fully to energy. For example, Hiroshima was eradicated by an atomic bomb that drew its energy from a quantity of matter weighing less than a penny. A cup of water completely converted to energy would provide enough electricity to power a million homes for a year. These are impressive figures. As usual, however, heavenly bodies put our Earthly comparisons to shame. Our Sun is a run-of-the-mill astronomical object as far as energy goes. Even so, every second it converts a mass of hydrogen equal to 15 billion cups of water into energy. That 1-second output represents more energy than humans have used in the history of civilization.

By the LIGHT of a Star

Each type of energy teaches us something different about the universe. However, our best tool for probing the cosmos is electromagnetic energy. Particles of electromagnetic energy called photons transmit light across the universe. From those signals, faint as they may be, we extract information about the motions of objects, how far away they are, their compositions and ages, and more. We owe our understanding of the universe to photons that have journeyed for billions of years across space. We snare these tireless travelers with giant telescopes or other surrogate eyes that we use to stare at the heavens.

There's much more to light than meets the eye. Electromagnetic energy spans a broad spectrum from high-energy gamma rays at one end to low-energy radio waves at the other. The visible light that humans use to perceive the world falls somewhere in the middle. We have a natural bias toward visible light and the colors of the rainbow that compose it. But visible light opens such a narrow window on our universe that we are practically blind to its wonders. We might as well try to drive on the highway with a windshield painted black except for a half-inch slit in the middle. Fortunately,

The Electromagnetic Spectrum

On either side of the familiar (but tiny) visible rainbow of red, orange, yellow, green, blue, indigo, and violet lie vast bands of nonvisible radiation. The higher the energy of the radiation, the shorter its wavelength. The spectrum divisions between the different forms of radiation overlap because the names derive in part from how the radiation is generated and the technology used to detect it.

Frequency (in hertz) | Wavelength (in centimeters)

10^{25} 10^{-15}

Gamma rays

high frequency short wavelength 10^{20} 10^{-10}

X-rays

Ultraviolet light

10^{15} VISIBLE LIGHT 10^{-5}

Infrared rays

10^{10} Radar waves   1
Microwaves
Television and FM radio waves
AM radio waves
10^{5} 10^{5}

low frequency long wavelength   1 10^{10}

astronomers have devised ways to expand this limited vision. Through each new window in the electromagnetic spectrum, the universe displays a different face.

Gamma rays carry the most energy of any photons. We rarely see gamma rays arise naturally on Earth. Certain types of radioactive decay produce them, and some energetic thunderstorms spit out gamma rays with their lightning bolts. Otherwise, they come from places in space where matter undergoes sudden and violent changes in form.

One such setting is the core of our Sun. The thermonuclear reactions that power stars make more gamma rays than all other kinds of energy. However, those photons must travel for tens of thousands of years to escape the Sun's interior. Along the way, they ricochet off countless atoms reeling like drunks down a crowded street. The collisions sap energy from the gamma rays and convert each one into thousands of lower-energy photons: x-rays, ultraviolet light, and visible light. By the time the photons emerge from the Sun, few gamma rays remain. Gamma rays also pour forth in unimaginably energetic bursts from mysterious objects near the fringes of the universe. Discovering how that happens remains an exciting challenge in astronomy.

X-rays are familiar to anyone who has broken a bone or gone to a dentist. They are less energetic than gamma rays but are strong enough to penetrate our flesh and most nonmetallic substances. In space, x-rays stream from the sites of stellar storms

Different Light Waves, Same Object

Astrophysicists learn different things about cosmic phenomena by studying them at different wavelengths of light. As shown here with the supernova remnant known as Cassiopeia A, the energy of the light emitted by various parts of the remnant reveals the physical processes occurring there, not all of which are apparent in a familiar optical image. In the x-ray image at right, for example, one of the first images from the Chandra X-ray Observatory, scientists discern two shock waves. Cas A was named in 1948 by radio astronomers who "rediscovered" it as the strongest radio source in the constellation of Cassiopeia. Optical astronomers later found faint wisps at the same location (*below*) and determined that Cas A is the remnant of an explosion that occurred about 300 years ago.

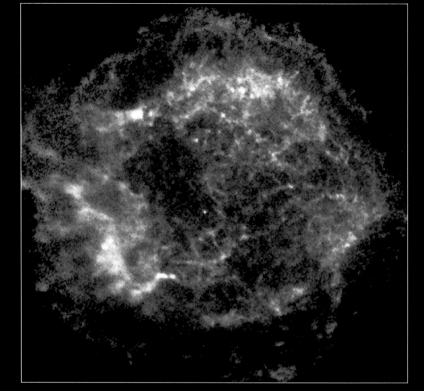

X-ray

Visible here are shock waves that arise as ejected material from the supernova explosion plows into space and encounters bands of gas and dust. The bright point near the center may be the neutron star or hot material around the black hole remaining after the explosion that produced Cas A.

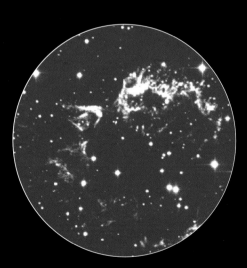

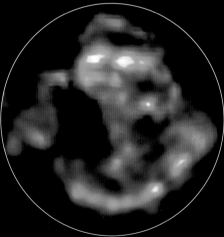

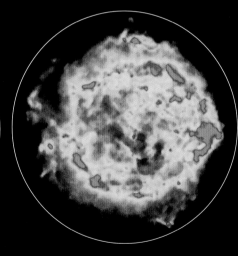

Optical

Matter with a temperature of about 10,000 degrees shows up at visible wavelengths. Some of these wisps are concentrations of heavy elements, dense clumps of ejected stellar material.

Infrared

Dust grains that have been swept up and heated to several hundred degrees by the expanding hot gas emit infrared radiation.

Radio

Long-wavelength radio emission comes from high-energy electrons moving in large spirals around magnetic field lines of force.

and explosions, including the solar flares that periodically erupt at active regions on the Sun. They also signal the presence of energetic but invisible objects such as neutron stars or black holes. Gas that wanders near one of these compact objects spirals into a whirling disk. Intense frictional forces make temperatures within the disk soar to millions of degrees as the gas funnels into the object. The hot gas radiates x-rays into space before it strikes the surface of the neutron star or, in the case of the black hole, vanishes from sight.

X-ray astronomy also reveals more mass in the universe than the stars alone expose. X-rays flow copiously from clusters of hundreds of galaxies. The photons come not from the galaxies themselves but from diffuse gas that fills and cocoons the clusters. This blazingly hot gas, up to 100 million degrees, exists within most thickly populated clusters of galaxies. The gas weighs about as much as all the stars in the universe combined.

Ultraviolet (UV) light lies just beyond the range of our eyes' sensitivity, although honeybees and many other insects see it quite well. Even though we can't see UV light, we encounter it every day as the most energetic light that reaches Earth's surface from the Sun. The ozone layer in the atmosphere absorbs most UV light, but enough gets through to cook any skin that's left unprotected by sunscreen. This type of light also breaks down molecules in plants and animals. Dwindling levels of ozone over parts of the globe may cause higher rates of skin cancer and other harmful biological effects.

The hottest stars, young monsters called blue giants, shine most energetically in UV light. These short-lived stars blaze thousands of times more luminously than our Sun. When blue giants clump closely together in space, they betray regions of vigorous star formation in galaxies. The Orion Nebula is one such hotbed of starbirth. Blue giants often remain together in beautiful clusters until they consume their nuclear fuel within a few hundred million years. The most obvious such cluster is the Pleiades, a lovely grouping of hundreds of stars that we also call the Seven Sisters.

Visible light falls within a narrow band, as we have noted. Within that band the colors that compose "white" light also have a range of energies. From least to most energetic, we know the colors as red, orange, yellow, green, blue, indigo, and violet. Students still learn a classic acronym for the order of the colors: "Roy G. Biv." However, you won't find indigo in most crayon boxes.

Greenhouses on Planetary Scales

Gases such as water vapor, carbon dioxide, and methane can form an insulating layer in the atmospheres of planets. Such a layer allows visible light to penetrate easily but would be more opaque to infrared radiation, trapping the heat near the planet's surface. Since this resembles what happens in a gardener's greenhouse—the glass walls and roof let visible light pass through but trap the warmed air inside—we call it the greenhouse effect. This phenomenon has been good to Earth: Without it our world would be on average many degrees cooler, possibly just enough to prevent life from forming here the way it did. Other worlds in our solar system have been less fortunate. Closer to the Sun than Earth is, Venus has suffered a "run-away" greenhouse effect, as shown on the opposite page.

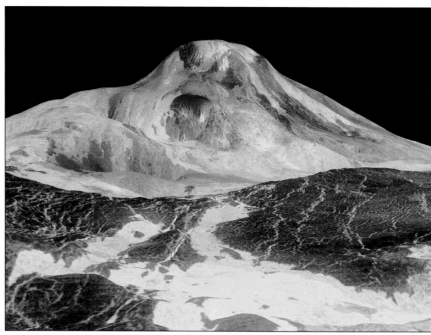

Fractured plains and the remnants of vast lava flows surround Maat Mons, a 5-mile-high extinct volcano on Venus. Vertical scale is exaggerated 22.5 times in this 3-D map, created from radar data collected in 1990 by NASA's *Magellan* orbiter. The simulated orange hues are based on color images from two Soviet missions to Venus in the 1980s.

The Sun shines most brightly in visible light. Biologists believe this may be why our eyes evolved to perceive only that range of electromagnetic energy. The Sun is a garden-variety star, so the energy from many other stars also peaks in or near visible light. For centuries visible light was all astronomers studied. They didn't have the technology to gather other forms of light. Visible light provides a reasonably accurate approximation of the universe because stars provide the bulk of its illumination. If we can see stars in visible light, we can see most of what's bright out there. We also see a little bit of just about everything else, since nearly all objects that glow emit some of their energy at visible wavelengths of light.

Like UV light, infrared light also just eludes our vision. We commonly think of infrared light as heat and sense it with our skin as warmth. Consider the heating element on your stove. Even after you turn it off and the red glow fades away, you can still feel its heat and you know not to touch it. Every warm object in the universe, including your body at this moment, emits infrared light. Special night-vision goggles can reveal this eerily glowing world to us.

All stars produce infrared light as well as visible light. But some relatively "cool" objects shine very faintly in visible light, so infrared offers the best way to detect them. These include the tiniest stars, which are the most numerous in the universe. They

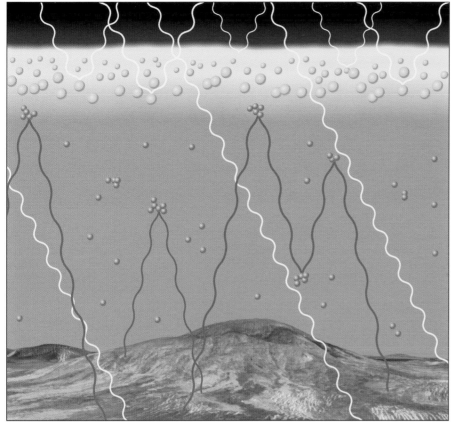

Cooked by the Sun, crustal rocks on Venus released quantities of greenhouse gases into the atmosphere. Fully 96 percent of the Venusian atmosphere is a choking shroud of carbon dioxide. As shown at left, infrared radiation (*red*) rising from the superhot surface of the planet is largely absorbed and reemitted back toward the surface in an infernal feedback loop. This "runaway" greenhouse effect finally stopped when the surface temperature reached about 900 degrees Fahrenheit, hot enough to vaporize lead or to bake a pizza in under 10 seconds.

will live for tens of billions of years after the Sun has used up its fuel. Still smaller objects, called brown dwarfs, are failed stars that never quite ignited nuclear fusion in their cores. They glow faintly with energy left over from their gravitational collapse. Planets also emit infrared light from their own internal heat. We may spot planets directly in other solar systems with infrared telescopes of the future.

Infrared light also figures in a notorious phenomenon of the day: the greenhouse effect. Visible light from the Sun bathes Earth and warms the surface. The land and sea then reradiate much of this energy into the atmosphere as infrared light. Certain gases—most notably water vapor, carbon dioxide, and methane—act like the glass panes in a greenhouse and prevent the heat from escaping into space. This effect is natural and not necessarily bad. Without it our planet might have been too cold for life to arise. However, during the twentieth century the exhaust from our factories and cars has stoked carbon dioxide in the atmosphere to ever-increasing levels. By trapping more heat, this extra gas may make the globe's average temperature rise by several degrees in the near future.

The effects of such rapid changes are uncertain. But we know that once a delicate thermal balance is perturbed, it may never recover. Our neighbor Venus is a case in point (*above*). That planet never shed the carbon dioxide atmosphere it had at birth.

Temperatures slowly rose, baking more carbon dioxide out of rocks near the surface. At some point a "runaway" greenhouse effect took over. It didn't stop until the surface reached a scalding 900 degrees Fahrenheit—hot enough to vaporize lead. Today, the atmosphere of Venus is almost 100 times denser than that of Earth and consists of 96 percent carbon dioxide.

Microwave radiation, slightly less energetic than infrared, is most familiar as a means to rejuvenate last night's casserole. A microwave oven heats the water molecules within all food. The molecules jiggle and vibrate, releasing their newly acquired microwave energy in the form of kinetic energy. Water is common in space as well, and it behaves the same way as it does in your kitchen. In clouds of cool gas, billions of tons of water vapor absorb microwaves emitted by nearby stars. Under certain conditions, charged gases around the clouds can stimulate the water molecules to emit microwaves as well. These microwaves are in turn absorbed and reemitted by molecules in the cloud, creating a powerful feedback loop. When this cycle becomes especially intense, beams of microwave radiation shoot into space. This curious event is called "microwave amplification by stimulated emission of radiation," or "maser" for short. Astronomers see these microwave outbursts all over the sky. If the word "maser" sounds familiar, it should. "Laser" is nearly the same acronym, with the word "microwave" replaced by

Stimulating Light

Light waves with lengths ranging from the thickness of a dime to that of a dime-store novel are called microwaves. In nature, microwaves are emitted by clouds of interstellar gas and also by the gaseous shrouds of newly formed stars. These interstellar clouds are largely made up of diatomic hydrogen, two atoms of hydrogen joined in a single molecule. When the clouds are surrounded by energetic plasma or radiation fields, the water molecules in the cloud can be stimulated to emit microwaves as well. A feedback loop can result (*right*), producing powerful beams of microwave radiation shooting into space. This phenomenon is called "microwave amplification by stimulated emission of radiation," or "maser" for short.

In interstellar space, a diatomic molecule of hydrogen (*red*), collides with a water molecule, shown here with a blue oxygen atom and red hydrogen atoms.

The collision transfers energy to the water molecule, causing it to jump to a higher energy state. The hydrogen molecule loses energy in the process.

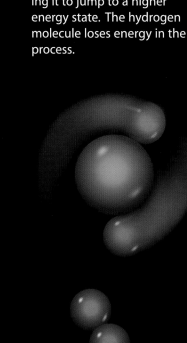

"light." Your CD player and pocket laser pointer do the same thing as a maser, except they amplify beams of visible light (usually red) instead of microwaves. If you had microwave vision, you'd also see pulses of radar from your friendly highway patrol officer and blazes of light from the relay towers used to transmit cellular telephone calls.

Microwave astronomy provides the most compelling evidence of our origins. Theory suggests that the universe was born in a Big Bang. If so, then its early searing heat—with temperatures that reached trillions of degrees—should have wafted evenly through space as the universe expanded. After billions of years, space should still be awash with the residual warmth, although at very low levels—less than three degrees above absolute zero, the coldest possible temperature. In 1965 the physicists Arno Penzias and Robert Wilson found this "cosmic microwave background radiation." The radiation has its peak strength in microwaves, but it also contains radio waves. Penzias and Wilson didn't even plan their observation, but it turned into the most important single discovery in the history of cosmology. Other lines of evidence also support the Big Bang, but microwaves are the strongest link. The next time you hear static between your AM radio stations at night, keep in mind that you are picking up a few cosmic microwaves from every direction in the universe—radiation that dates from the beginning of everything we know.

When the excited water molecule drops to a lower energy level, it emits a photon (*yellow*). Rather than dropping to the lowest ground state, the molecule shown here has dropped to an intermediate metastable state, where it can remain for some time.

When a passing photon strips the metastable molecule of its energy, it drops to the ground state and emits a photon of the same wavelength, traveling in the same direction as the incoming photon. On encountering other metastable molecules, the two photons set off a chain reaction of emissions.

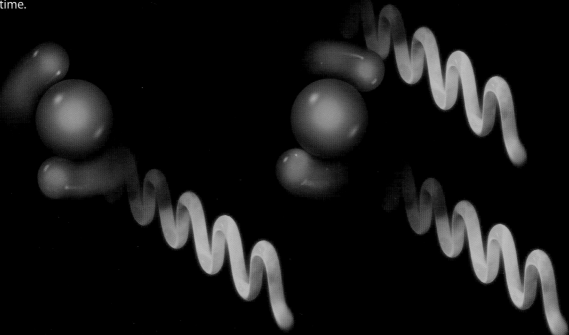

Radio waves contain the lowest-energy photons of all electromagnetic radiation. These photons penetrate through vast reaches of gas and dust to give us deep views of our galaxy and the universe. Because most atoms and molecules barely interact with radio waves, the daytime sky is just as transparent to radio waves as the night sky. Therefore, radio astronomers can work 24 hours a day. Their observations have shown that the center of the Milky Way in particular shines brightly in radio waves. So do quasars, named for a shortened version of "quasi-stellar radio sources." For years these bright beacons were mysterious. We now believe they shine from the cores of extremely active galaxies, where massive black holes gobble matter with abandon.

Ever since the Italian engineer Guglielmo Marconi invented the wireless radio, communication via radio waves has become central to our civilization. We hardly think twice about seeing and interacting with events on the other side of Earth, not to mention with astronauts orbiting in space and robots on the surface of Mars. Giant radio telescopes detect faint transmissions from spacecraft in the outer solar system even though their signals are weaker than one-billionth of a watt by the time they reach Earth. Indeed, radio waves are the most likely means for us to make the ultimate long-distance call: detection of intelligent life in other planetary systems.

All branches of astronomy, from radio to optical to gamma ray, share the need to collect as many photons of light as possible. The universe is a busy place when it comes to light. Signals from distant objects can vanish easily in the hubbub. Light pollution in the night sky on Earth doesn't help, nor does the din of our communications technology. It's the classic cocktail-party challenge on a cosmic scale: To listen to one "conversation" among many—an asteroid beyond Neptune or a single remote galaxy—we must amplify its light as much as possible while suppressing the rest.

Galileo was the first to do this. In 1609 he developed a 1-inch-wide telescope that made objects appear 30 times closer to Earth than they appeared to the unaided eye. Since then, astronomers have made steady progress in building ever-larger telescopes to collect visible light. The largest optical telescopes in the world today are the Keck I

At radio wavelengths, the spiral pattern of the classic Whirlpool Galaxy appears as clearly as it does at visible wavelengths. Astronomers speculate that the galaxy's spiral structure is primarily due to its gravitational interaction with the smaller galaxy that lies behind it (*right, top*), which creates a spiral density wave. Since the pattern also shows up in radio emission, the magnetic fields in the Whirlpool may also be compressed by the density wave.

Earth's Atmospheric Shield

Electromagnetic radiation bombards Earth from all directions in the universe, the bulk of it from our own Sun. Only certain wavelengths of radiation get through to the surface because Earth's atmosphere reflects, absorbs, or scatters the rest. Oxygen and nitrogen atoms in the thermosphere absorb nearly all x-rays and gamma rays, the most energetic forms of light; the mesosphere and stratosphere screen the remainder. A significant portion of ultraviolet light entering the mesosphere and stratosphere is absorbed by ozone molecules, protecting us from lethal doses. At 10 miles, radiation of longer wavelengths—from visible through radio—enters the troposphere, where water vapor, trace gases, carbon dioxide, dust particles, and pollutants absorb the infrared. Long-wavelength radio waves slide right past these small particles, however, making the radio window one of the most transparent for astronomical observation.

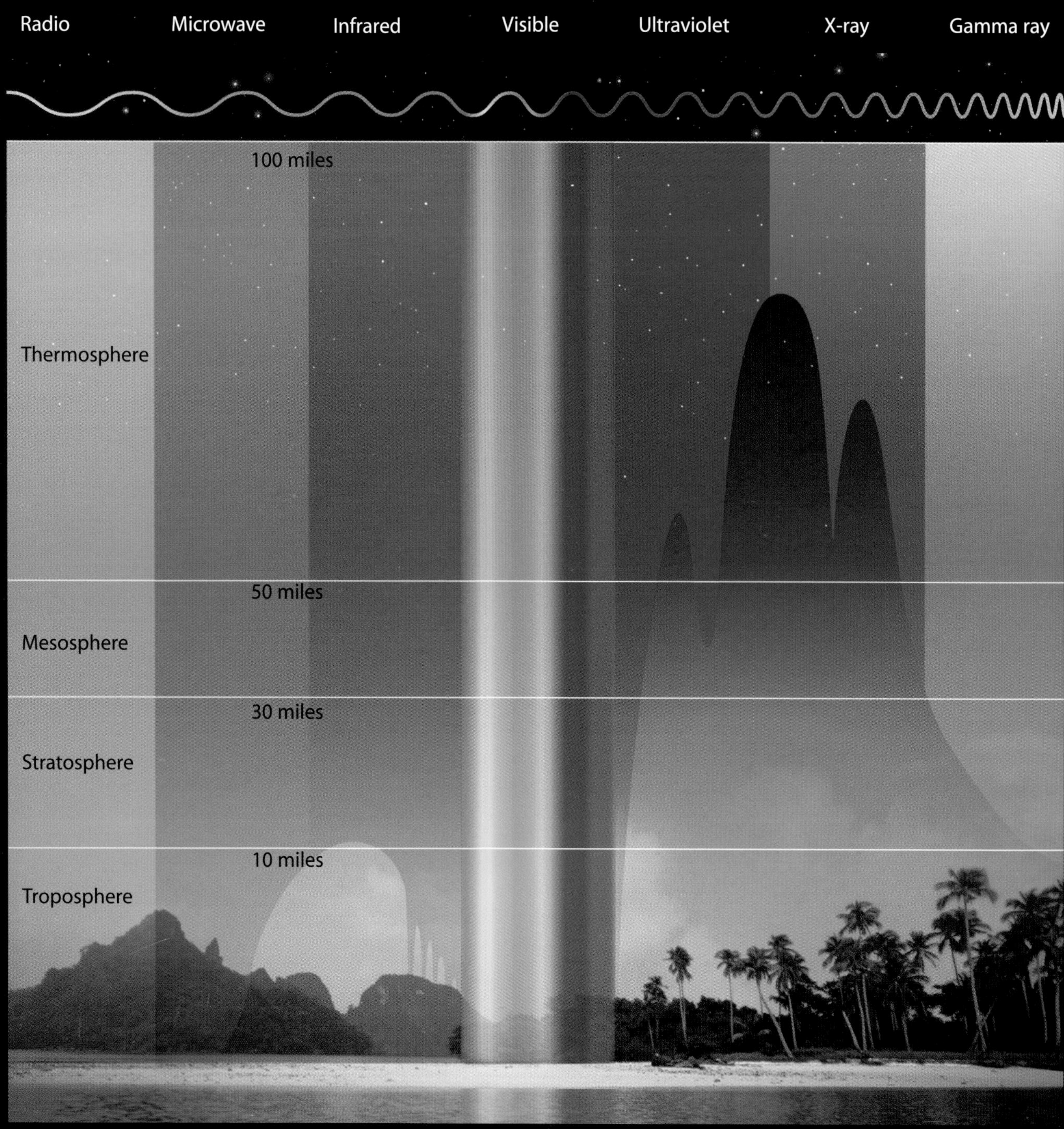

Radio · Microwave · Infrared · Visible · Ultraviolet · X-ray · Gamma ray

100 miles

Thermosphere

50 miles

Mesosphere

30 miles

Stratosphere

10 miles

Troposphere

and II telescopes in Hawaii. Both telescopes are over 33 feet across. Each mirror actually consists of 36 hexagonal segments, fitted together in a honeycomb pattern and held steady by hundreds of computer-controlled pistons. These marvels of engineering gather more light in 10 seconds than your eyes could see in three years without blinking. Other telescopes have single mirrors up to 27 feet across or multiple mirrors that provide even more light-gathering power. This fleet of observatories has ushered in a new golden era of optical astronomy.

We have also left the ground in pursuit of more photons. Telescopes in orbit around Earth offer two advantages. First, they let us detect electromagnetic energy that the atmosphere blocks from reaching the surface. This includes gamma rays, x-rays, ultraviolet light, and some infrared light. Second, they eliminate the blurring effects of Earth's blanket of air. The atmosphere is a fluid that sloshes back and forth like water in a pool. Imagine lying on the bottom of a pool and trying to read a sign held above the surface. You could do it if the water was perfectly still, but a tiny splash turns the letters into a muddled mess. By flying above the atmosphere's currents, the Hubble Space Telescope and other orbiting observatories see stars as crisp points rather than twinkling blurs.

Astronomers have devised a way to erase most of this distortion for ground-based telescopes as well. This still-developing technology, called adaptive optics, is a key observing strategy of the Keck telescopes and at other big observatories. In one version of adaptive optics, a laser next to the telescope beams high into the atmosphere to create a small artificial "star." The telescope aims at the laser star, which twinkles just like a real star—thanks to the shifting column of air above the telescope. A computer analyzes this pattern of light and calculates how to cancel out the distortions. Then, a thin mirror inside or next to the telescope carries out those calculations by deforming several times per second. The mirror's rippling shape counteracts the atmosphere's ebb and flow. By the time the light bounces off that mirror and into the observing instruments, it's as if the atmosphere was barely there at all.

Another way to get more bang for the telescopic buck is to combine light waves from two or more telescopes. The resulting images are sharper than any one telescope can yield. This technique, called interferometry, requires extensive mathematical analysis. Interferometry works best for radio telescopes because radio waves have the longest wavelengths and are the simplest types of light to combine accurately. Huge radio dishes in arrays that span many miles or even entire continents take exquisitely detailed images of the centers of galaxies and other distant objects. Satellites orbiting Earth use a similar technique with radar waves to gauge subtle shifts in land elevation. The satellites can spot the strain building along earthquake faults, as well as bulges in volcanoes that foretell eruptions.

Improvements in the way we record light have gone hand in hand with advances in telescope design. In the mid-nineteenth century, the American astronomer Henry Draper was the first to mount photographic plates on the back of a telescope. This was a leap forward from drawings and written notes, which depended on the observer's skill and accuracy. To this day, photographs produce stunning views of the heavens. For research, however, most astronomers now use digital detectors called charge-coupled devices, or CCDs. CCDs are silicon chips that convert most of the photons striking them into electronic pictures. That's much more efficient than the measly few percent conversion rate of film. CCDs also erase quickly to record the next image—no reloading necessary—and produce files that store easily. If you own a video camera, an inexpensive version of this device is your camera's electronic retina.

Not all observatories use mirrors or giant dishes to detect energy. Gamma-ray telescopes in space intercept their quarry with clear crystals, which then emit flashes of visible light. Similar flashes occur within the huge tanks of water at neutrino observatories like the ones that detected the neutrinos from Supernova 1987A. Astronomers also have lowered long strings of light sensors beneath the ice of Antarctica to detect neutrinos that stop there. A successful neutrino "telescope" would allow a direct view of the Sun's core, which flings out more than 100 trillion trillion trillion neutrinos every second. But the challenge to see neutrinos efficiently is steep indeed. After all, they can penetrate a light-year of solid lead (nearly 6 trillion miles) without effort.

Yet another nonelectromagnetic window on the universe may open soon: gravitational waves. According to Einstein's general theory of relativity, these elusive ripples in the fabric of space-time spread outward from the motions of massive objects. Two black holes spiraling around each other, or two neutron stars colliding, would stir space-time like an egg beater. As the gravitational waves passed Earth, they would alternately compress and expand the space between two objects. We would hardly notice the change, though. Objects separated by several miles would move relative to each other by less than the diameter of a single proton.

Spurred by the challenge of detecting such tiny motions, physicists are building gravitational wave observatories in the United States and Italy. At each of the two U.S. sites, pairs of mirrors hang 2.5 miles apart along both arms of an L-shaped tunnel. Laser beams course through vacuum tubes between the mirrors to measure their positions. Plans call for a similar observatory in deep space but with arms spanning millions of miles on a side. Ultimately, this approach may let us perceive the subtle ripples in space-time that reverberate through the universe from the Big Bang itself.

Probing Space with SPECTRA

Gravitational waves would provide a clever way to "see" the universe without photographs. Even so, beautiful pictures of heavenly objects will continue to inspire awe and wonder about the cosmos. There's something extraordinary about seeing rocks on the surface of another planet, the nurseries of new stars, or pinwheels of a hundred billion suns like our own. Astronomers enjoy such pictures as well as learn from them. But they have another tool to help unveil the scientific details about distant objects: spectra, which can be worth a thousand pictures.

Spectra are the patterns created by spreading the light from a glowing object into its component colors. Shining sunlight through a prism is an easy way to see this effect. Isaac Newton was the first to use a prism in this way. He also showed that a second prism, inverted with respect to the first one, recombined the colors of the spectrum into the original beam of sunlight. Rainbows are spectra of sunlight painted against the sky. Millions of raindrops act like spherical prisms. They refract and reflect sunlight back to your eyes at a precise but slightly different angle for each color. The geometry of this bending process creates a circle of brilliant colors. However, we usually only see the top halves of the circles because the ground gets in the way.

The difference between these spectra and the ones produced by telescopes is a matter of detail. Instruments called spectrographs divide the light of stars or galaxies into sharply focused spectra that spread up to several feet long. These spectra are not typically smooth transitions from one of the "Roy G. Biv" colors to the next. Rather, fine forests of bright and dark lines may cut across the rainbow, looking like rows of supermarket bar codes. These features contain a wealth of information about cosmic objects that we cannot learn in any other way.

To understand why these lines exist, we must consider the fundamental nature of light. The German physicist Max Planck showed in 1900 that light emitted by a hot object must radiate in tiny discrete units, which he called quanta. "Quantum mechanics" comes from this name. Einstein further demonstrated the need for quanta in 1905, proving that light is packaged in quanta that travel in waves. The notion that light travels simultaneously as a particle and a wave was one of the first signs that something was odd in the subatomic realm.

In 1913, the Danish physicist Niels Bohr proposed the idea of a quantized atom: a system where the electrons orbiting the nucleus could exist only at specific distances and thus only at specific levels of energy. This is akin to being forced to hop down a staircase two or three steps at a time. The theory of quantum mechanics—developed in detail by many physicists in the decades after Bohr's proposal—places explicit limits on the motions and energies of electrons in excited atoms. It's hard to understand why these limits should exist, but as strange as they are, they explain so many aspects of our world perfectly. You might be surprised to learn that the luminous hands on your wristwatch and the glowing stars on the bedroom ceilings of many children are driven by quantum mechanics. Bright light pumps up the energy levels of the electrons in the special materials. Then, when you turn off the lights, the electrons cascade back down. They release photons with a specific energy: that familiar green glow.

More important for astronomy, the features we see in cosmic spectra depend on quantum mechanics. When electrons in energized atoms bounce from one energy level to another, they absorb or emit photons of light that contain those precise quantities of energy. If enough atoms absorb photons of a particular energy, that part of the spectrum exhibits a dark line. The atoms have swallowed the light. Elements in the atmosphere of a star can play that role. Light from inside the star passes through the elements on its way to Earth. Electrons in the elements subtract light at specific places in the spectrum, leaving dark gaps. Conversely, when enough atoms emit photons of one energy, a bright line appears at that spot in the spectrum. The ensemble

Decoding Spectral Lines

All spectra are the result of specific kinds of discrete energy transactions, gains or losses of energy when an atom is excited and an electron is disturbed from its orbit around the nucleus. Of the three types of spectra, the most familiar is the continuous spectrum. This rainbow band of color is produced by very crowded, energetic conditions, such as in the depths of a star, where the distinctions between energy levels gets smeared. As explained below, the two other types—absorption and emission spectra—are produced by electrons jumping or falling from one energy level to another.

Electron at a higher energy level

Absorption

When an electron absorbs a passing high-energy photon, it jumps to a higher energy level at a greater distance from the nucleus, represented here by an expanding probability cloud. Many such interactions create gaps in the continuous spectrum. The total effect for all the atoms of a given chemical element is a distinctive pattern of absorbed wavelengths such as the simplified absorption spectrum for sodium at right.

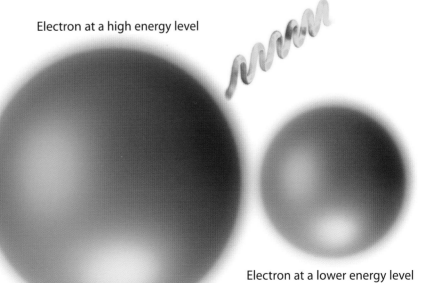

Electron at a low energy level

Sodium absorption spectrum

Emission

Electrons kicked to higher energy levels will quickly drop back down, giving up a photon whose wavelength matches the energy difference between the two levels. In the case of sodium these emissions produce the characteristic yellow light of sodium streetlamps (*above*) and the sodium cloud around Jupiter's moon Io (*below*).

Electron at a high energy level

Electron at a lower energy level

Sodium emission spectrum

Soon after astronomers first used spectroscopy, they discovered one-quarter of the visible universe.

of energy levels for each element is unique. Scientists have charted these spectral "fingerprints" in their laboratories for decades by studying patterns of light from hot and cool gases. Because many elements coexist within gases and other objects in space, each spectrum can be a collage of dozens of elemental fingerprints. The same technique can be applied to molecules such as water vapor or carbon monoxide. Those assemblages of atoms also leave characteristic imprints on spectra.

To picture in another way how this works, think back to the last fireworks display you saw. Exploding rockets lit the night with incandescent blues, reds, and other colors. Pyrotechnicians burn different materials for each color—magnesium for white, iron for red, copper for green, and so on. Now imagine what happens when those elements vaporize on the hot surface of a distant star. They produce colors in exactly the same way. When we take a photograph of the star, the colors blend together like paints on a long-unwashed palette. However, when we measure the spectrum of the star, the light spreads into its components and reveals the contributions of each element and compound.

Soon after astronomers first used spectroscopy, they discovered one-quarter of the visible universe. In 1868 the first detailed spectrum of the Sun's atmosphere exposed many familiar elements, such as hydrogen and oxygen. But no known atoms of the day could produce other features that arose. One set of spectral features was so prevalent that astronomers attributed them to a new element. Since it first appeared in the Sun—known to the Greeks as Helios—the element was dubbed helium. Today, we know that almost 25 percent of the atomic matter in the universe is helium. Here on Earth we encounter it in party balloons and Super Bowl blimps. Planetary atmospheres, the surfaces of asteroids and comets, the remnants of supernova explosions, and other objects reveal their unique mixtures of ingredients through their spectra in the same way.

Beyond composition, spectra contain a storehouse of other astrophysical data. One quantity we can deduce is temperature. Each element and compound burns at a certain temperature. For example, you could judge the temperature in your kitchen oven without a thermometer by tossing some junk mail inside. When that useless paper bursts into flames, you know the oven has crossed the magic threshold made famous by Ray Bradbury's science-fiction novel *Fahrenheit 451*. In a similar way, when we see a particular set of emission lines from a glowing object, we know that the object's temperature is at least as high as the burning point of that substance. Absorption lines also reveal temperatures. A cool star emits photons with relatively low energies. Certain atoms and molecules in the star's atmosphere absorb light at just those levels. If the star's temperature increases, the electrons jump to higher levels. We then see a different set of dark lines across the spectrum.

The strength of an object's magnetic field is another characteristic we can learn from a spectrum. Some spectral lines split into two parts in the presence of a magnetic field. As the field grows stronger, more lines show this effect. We also gauge field strength by observing the motions of nearby charged gas, which a magnetic field steers through space. However, spectral splitting is our most direct probe.

Finally, spectra have proven invaluable as tracers of the motions of objects. Spectral lines are not static. Rather, they shift when the glowing object moves toward or away from us. Our Earthly analogy is the Doppler effect (page 139). The pitch of an approaching race car sounds higher than it actually is, while one receding from you sounds lower. In the case of light, its wavelength—a property directly related to its energy—undergoes a "blue shift" to shorter wavelengths (higher energies) if the object approaches. A receding object displays a "red shift" to longer wavelengths of light (lower energies). The entire spectrum shifts, but the easiest changes to spot are the shifted positions of the object's bright emission lines and dark absorption lines. Even so, these changes are subtle. Unless the object moves at a fair fraction of the speed of light, we must use precise instruments to measure the shifts.

The rewards are well worth the effort. For example, when a star spins, half of it turns away from you while the other half turns toward you. As a result, part of each spectral line has a slightly shorter wavelength, while the other part has a slightly longer one. The net effect is that each line gets wider. The faster the star spins, the wider its lines become. The Doppler effect works on larger objects as well, no matter how far away they are. It shows us that spiral galaxies, millions of light-years away, spin on their axes at more than 100 miles per second. Even at that clip our gigantic Milky Way takes more than 200 million years to go around once. Moreover, when galaxies reside in rich clusters containing hundreds or thousands of similar galaxies, they dart around within those clusters at speeds up to 1,000 miles per second.

Edwin Hubble tracked red shifts in the spectral lines of his galaxies to determine that the entire universe is expanding. Here, our Doppler shift analogy suffers a bit. Distant galaxies are not roaring away from us like the race car in our Earthbound comparison. Rather, all of space expands, and galaxies get carried along for the ride. The relentless expansion stretches a galaxy's light waves to longer and longer wavelengths as the waves travel toward Earth. Consequently, their spectral lines are redshifted just as though the galaxies are blasting through space at high speeds. Galaxies near the limits of the observable universe appear to recede so quickly from us that many of their spectral lines disappear from the spectrum of visible light. They shift all the way into the infrared.

Extremely tiny Doppler shifts in the spectra of stars are the key tools to finding planets outside our solar system. A star does not sit motionless at the center of its planetary kingdom. Rather, a planet's gravitational pull tugs its star to and fro—just as when you jump into the air, you induce Earth to move slightly in the opposite direction. Both phenomena are displays of Newton's third law: For every action there is an equal and opposite reaction. Thus, a star with planets in orbit around it jiggles slightly in space, leading to minuscule but periodic red shifts and blue shifts in the star's spectral lines. Starting in the mid-1990s, astronomers used ultrasensitive spectrographs to search for such jiggles in other stars. Current technology allows them only to detect planets close to or exceeding Jupiter in size. However, even that high limit has yielded more planets outside our solar system than inside it. New search programs will extend this quest down to planets with masses like that of Saturn, then Uranus and Neptune. Within a decade or two, planets like Earth may be within range. As tempting as it was for us to think of our solar system as something special, it's already clear that planets in the cosmos are far from rare.

ELECTROMAGNETISM at Work

The light from distant galaxies and from stars within our own galaxy is a form of electromagnetism, one of nature's four basic forces. Other types of electromagnetic energy are all around us, although they may seem radically different from starlight. For instance, the electricity that powers our society arises from the same force that governs light. So too does the magnetism that guides sailors and scouts and encodes the strips

The Doppler Effect: Key to Motion and Energy

Astronomers use the absorption and emission lines of spectra of celestial bodies as a key to their motions and energy levels, clues to unveiling the nature of even the most distant galaxies in the sky. Scientists can do this because of the so-called Doppler effect. The sound emitted by a race car has a higher pitch as the car approaches because the wavelengths of sound are compressing in the direction of the observers (*top*). A race car moving away has a lower-pitched sound because the intervals between waves stretch. The same principle applies to electromagnetic radiation and is the key to reading the spectra of stars and other celestial objects. By analyzing the location on the spectrum of certain patterns of spectral lines (*bottom*), scientists can determine whether an object is moving toward or away from Earth and how fast it is moving. This information in turn figures into calculations of the object's mass and the kind of energy it is emitting. Added together, these separate clues can suggest, for example, that a powerful source of radio waves harbors a supermassive black hole (page152) or that a star 50 light-years away is actually home to a small system of planets (page 140).

The Doppler Effect and Red Shift

Absorption lines for a particular chemical element appear at a certain place on the spectrum for an object at rest in relation to an observer (*left, top*). (The lines used here have been simplified for clarity.) If the object is moving away from the observer (*middle*), the wavelengths stretch; some lines move off into the infrared, while lines normally in the blue range shift toward red. For an object moving toward the observer, the wavelengths compress and lines shift toward the blue end of the spectrum (*bottom*). The spectra of rotating objects show both blue and red shifting (page 155).

Detecting Extrasolar Planets

Careful measurements of tiny shifts in spectral lines—the result of the Doppler effect—allow scientists to identify stars that might be home to planetary systems. In the case of Upsilon Andromedae, variations in the velocity of the star's rotation (*graph at right*) strongly suggest that a trio of planets orbits this star, which is in the constellation Andromeda.

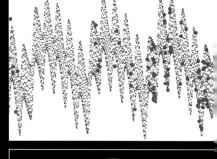

The smallest body, at three-fourths the mass of Jupiter, orbits so near the star that its "year" is only 4.6 days long. The gravitational influence of this sidekick causes the star to buzz up and down in the envelope of its longer-period cycles. The middle planet, twice the mass of Jupiter, causes the star to wobble with a period of about 242 days. The outermost planet, at about four times Jupiter's mass, lumbers around the star in about 3.5 years. The presumed orbits are plotted at right, with the orbit of Earth (*dotted line*) included for reference.

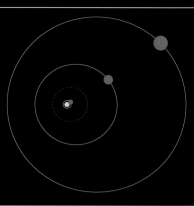

Below is an artist's rendition of what this extrasolar system might look like from behind the outermost planet.

on our ATM cards. The links among these seemingly diverse phenomena extend electromagnetism's reach into every niche of the cosmos and every part of our lives.

Those links took time to unveil. In 1820, the Danish physicist Hans Christian Oersted first demonstrated that an electric current flowing through a wire made a magnetized compass needle swing around. That is, a moving electric current created a magnetic field. A decade later, the English physicist Michael Faraday proved the reverse: A magnet moving through a coiled wire induced an electric current in the wire. This pair of results demonstrated the inextricable ties between electricity and magnetism. Not until the 1860s, however, did anyone make the leap to add light to that pairing. The Scottish physicist James Clerk Maxwell theorized that light waves stem from fluctuating electric fields and magnetic fields that stream together through space. Changes in the magnetic field generated a changing electric field, and vice versa, as Oersted's and Faraday's research showed. These mutual fields, oriented perpendicular to each other, moved at the speed of light: 186,282 miles per second. According to Maxwell's equations, electromagnetic waves moved in the same way as the types of periodic waves that travel along a snapping rope or the rippling surface of a pond.

Maxwell's work cemented the relationship of light, electricity, and magnetism. Other than the omnipresent waves of light around us, electricity is the most apparent of these in our daily lives. The carrying units of electricity are electrons, the negatively charged particles so important in quantum mechanics. Free electrons are tiny. A billion trillion of them weigh just a millionth as much as a paper clip. This makes the power of their electrical charge all the more impressive. Ounce for ounce, electromagnetic power is 100 billion times stronger than the weak nuclear force and 40 orders of magnitude stronger than gravity. For example, the electrical needs of a four-bedroom home for an entire year require less than one-thirtieth of an ounce of electrons. If the electrons from 1 cubic centimeter of the Space Shuttle's nose cone were moved down to the launch pad, the extra electrical force at the pad would prevent the shuttle from lifting off. Just 2 pounds of electrons clumped together on the far side of the Moon would overwhelm Earth's gravitational pull and tear the Moon from its orbit. These things don't happen, however, because a positively charged proton exists for every negatively charged electron. It's nearly impossible to create a big surplus of either because nature hates an imbalance of electrical charge.

When too many electrons build up in one spot, the resulting forces can create a channel of electrically charged gas—a plasma. An electric current will immediately spark through the plasma to relieve the imbalance. You don't mind feeling that shock when you touch a car door or your dog's nose, but you'd prefer not to ignite a spark on your computer's circuit board. The strongest naturally occurring sparks are considerably more hazardous. Billions of tiny ice particles rub together to produce electron imbalances within storm clouds. When the electrons arc to another cloud or to the ground, we call the spectacular discharge lightning. Our salty fluids make our bodies good electrical conductors. As a result, plasma channels often connect to someone's head—or the upraised club of a careless golfer—to launch the discharge.

These fatal shocks on Earth pale next to those on Jupiter and Saturn, where thunderclouds are thousands of times larger. Lightning bolts from the gas giant planets would devastate entire cities if they struck here. The system of satellites around Jupiter hosts an even more impressive electrical display. The volcanic moon Io ejects sulfurous gas into space. This charged material circles Jupiter in an enormous torus— a doughnut-shaped region laced with magnetic fields. As Io orbits Jupiter and passes through the torus, a current flows between them. Carrying about 2 trillion watts, it's the most powerful electrical circuit in the solar system, save for those on the Sun. The current creates bright auroras near Jupiter's poles and makes the environment near Io dangerous for spacecraft.

Smaller electric currents course through Earth's interior and, to varying degrees, the interiors of other planets. Earth contains an iron core, as well as minerals near the surface that are susceptible to electromagnetic forces. Iron within the molten outer core flows readily, while slow stirrings in the overlying rock carry material toward and away from the core. This drives a weak but large-scale electric current. The exact mechanism still isn't clear, but, as Oersted showed, moving electric fields spawn

Contributing to the constant emission of sulfurous gas that fuels a powerful electrical circuit between Jupiter and its volcanic moon, two sulfurous eruptions are visible on Io in this color composite image from the *Galileo* spacecraft. On the far left, a bluish plume rises about 85 miles above the surface of a volcanic caldera called Pillan Patera. In the middle of the image, near the day/night shadow line, the ring-shaped Prometheus plume rises 45 miles above Io and casts a shadow to the right of the volcanic vent. The Prometheus plume has been visible in every image made of this region, dating back to the *Voyager* flybys of 1979, raising the possibility that this volcano has been continuously active for at least 20 years.

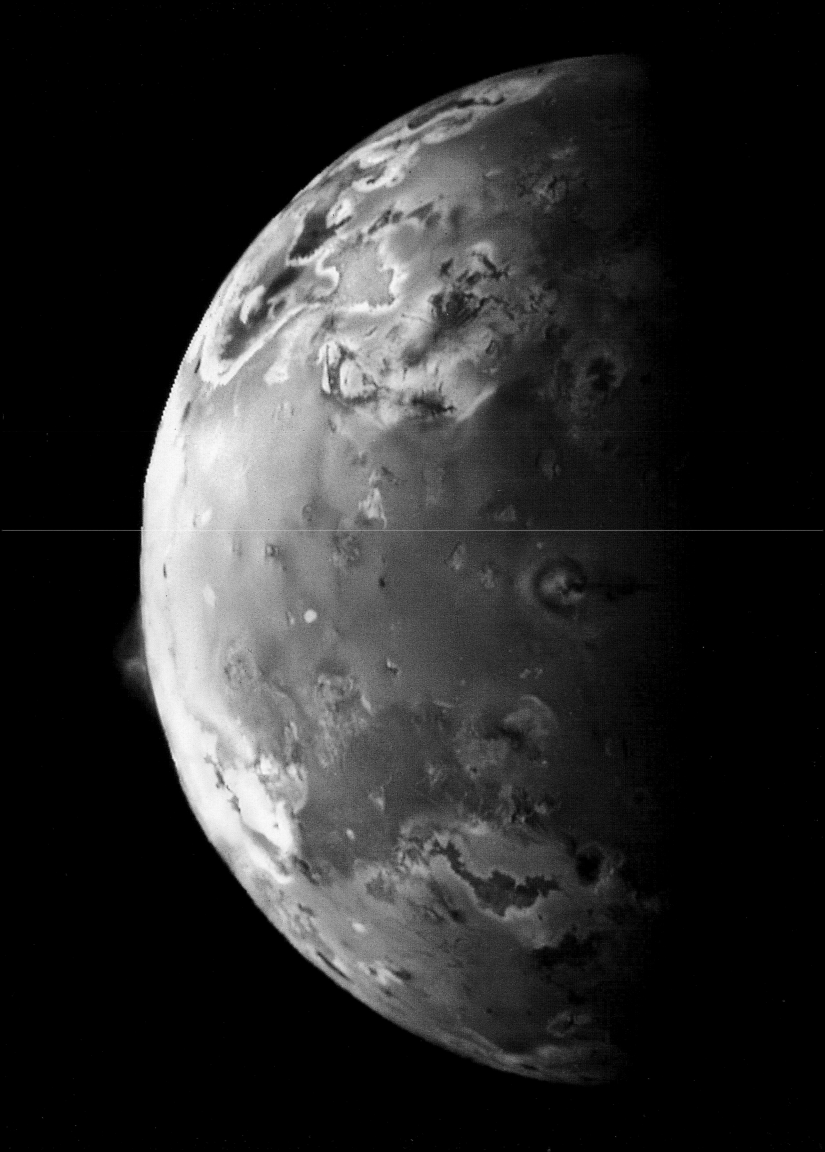

magnetic fields. Earth therefore generates a magnetic field like a giant bar magnet. The magnetic poles happen to lie near the geographic north and south poles of our planet, making compasses useful in navigation. That hasn't always been the case. Earth's magnetic field has weakened, wandered, and flipped many times throughout its history.

Earth's field at the surface is strong enough to drive a compass needle, but it is weak overall. Magnets on your refrigerator are hundreds of times stronger. Even so, Earth puffs a magnetic bubble into space—a "magnetosphere"—that is sturdy enough to shield us from a constant onslaught of charged particles from the Sun and deep space. The magnetosphere deflects some particles around Earth and traps others within belts of radiation that girdle the planet. Without this protective force field, the Sun's blasts would long ago have wiped out life on the planet's surface. The other planets also generate magnetospheres. It comes as no surprise that Jupiter, with its massive core of metallic hydrogen, generates the grandest one in the solar system. If we could see it glow in the night sky, it would appear several times larger than the full Moon.

Planetary magnetic fields are puny compared to those on the Sun. The gaseous outer layers of the Sun spin once every 25 days at the equator and every 31 days near the poles. But the center of the Sun spins like a solid ball. Between the outer and inner layers, astronomers have detected a narrow zone where the Sun's electrically charged gas shears, almost like clouds caught between slow and fast streams of air in Earth's atmosphere. This shearing action generates intense magnetic fields. The fields try to line up but quickly get distorted and fold over on themselves because of the Sun's differential rotation. The physical equations that govern the resulting stresses are related to those that describe tension in a twisted batch of rubber bands. Severely stretched field lines can abruptly snap through the Sun's surface, creating turbulent regions and sunspots. The magnetic turmoil ejects flares of charged particles and blobs of the Sun's atmosphere far into space, much as a naughty schoolchild uses rubber bands to propel wads of paper across the room. When these particles reach Earth, they can harm satellites and disrupt electronic communications. The Sun's particles also collide with air molecules in Earth's upper atmosphere. The molecules temporarily absorb energy and reemit it in the form of visible light. We see the results as auroras—colorful ribbons of light in the sky near the north and south poles.

Twisted arcs of magnetic force carry spectacular eruptions of solar matter into space. Without Earth's own protective magnetic shield, we would be exposed to lethal doses of high-energy radiation and solar particles.

Particularly energetic outbursts from the Sun can produce auroral displays over broad regions of the globe.

Sighting the SUPERENERGETIC

Elsewhere in the cosmos, magnetic fields lace our Milky Way and entire clusters of galaxies. These fields sweep through millions of light-years of space, guiding the motions of vast curtains of charged gas. They also accelerate protons and atomic nuclei that supernova explosions have spewed into space. These particles gain enormous amounts of energy during their looping travels around the galaxy. If they slam into Earth's atmosphere, we call them cosmic rays. Some of them carry so much energy that they have the mass equivalent of a brick. Cosmic rays can damage satellites and computer chips, and they make spurious blips of light on astronomical CCDs. They also account for about 20 percent of the natural radiation that we encounter at Earth's surface.

The most impressive magnetic fields of all may encase the bizarre objects called "magnetars." These neutron stars appear to have magnetic fields 100 times stronger than ordinary neutron stars and a billion times stronger than the strongest fields ever sustained in laboratories on Earth. Such intense fields may crack the exotic surfaces of the dense stars, generating bursts of gamma rays and x-rays that make them visible to us. The crushing force would bring the rapidly spinning stars to a near standstill within thousands of years, a cosmic eyeblink. Once the stars slow down, their radiation would dwindle away. Millions of these quiet objects may populate the Milky Way, utterly invisible to us.

Neutron stars are just a few pinches of stellar mass away from black holes. But the mass of a solitary black hole, as impressive as it is, comes nowhere near that of supermassive black holes, thought to dwell at the hearts of many galaxies. These behemoths may contain hundreds of millions of times more mass than our Sun, all crammed into a volume smaller than our solar system. We call such systems gravitational engines because they use their immense tidal forces to convert

Familiar to inhabitants of the Northern Hemisphere as the Northern Lights, the dancing sheets of excited atmospheric particles called auroras also occur in the Southern Hemisphere. Here they color the Australian sky an eerie red.

Jupiter's magnetosphere is the grandest one in the solar system, about 10 million miles wide and 500 million miles long. If we could see it glow in the night sky, as in this artist's rendition, it would appear several times larger than the full Moon.

No one quite knows how the gargantuan black holes form in the first place. But once they begin adding entire stars to their menu, their gravitational hunger only increases.

gravitational potential energy into superenergetic bursts of light. Their strategy is ruthless but simple: just rip all incoming matter to shreds.

Think back to James Joule's water-paddle experiment and his reasoning that some of the potential energy of water plunging over Niagara Falls should convert to heat because of the resistance of the water itself. In the case of a supermassive black hole, entire stars play the role of the falling water. Stars that wander too close to the black hole feel its tidal forces and are pulled apart. The spiraling gas forms a thick disk that whirls ever closer to the hole. The gas cannot just plummet into the black hole because the material in the disk is too dense. The disk fiercely resists free-falling motion, like the water within the falls. Therefore, the gravitational potential energy of the gas converts almost entirely to heat, and the disk shimmers with a temperature of millions of degrees. The infalling gas prevents gamma rays and x-rays from escaping out the sides of the disk. Instead, the energy blasts into space above and below the disk in colossal jets, carrying along matter at more than 99 percent the speed of light. The surrounding gas glows with visible light and radio waves that we can see to the edge of the observable universe.

This is the leading model for the incredible energies produced by quasars. No one quite knows how the gargantuan black holes form in the first place. But once they begin adding entire stars to their menu, their gravitational hunger only increases. In vigorous quasars the black holes eat as many as 10 stars a year. Less active galaxies can maintain their luminosities on more meager diets.

Most of this action in distant quasars took place long ago. We see none in our neck of the universe today. One perfectly reasonable explanation is that the nuclear regions of nearby galaxies have run out of stars to feed their engines. Perhaps the black holes already devoured all stars whose orbits came too close. No more food, no more explosive regurgitations.

There's also a more fascinating explanation. As a black hole's mass grows, its event horizon—the point of no return for matter and for light plunging into the hole—also expands. This has the counterintuitive effect of decreasing the black hole's tides. The critical factor in tidal forces is not the total gravity felt by a falling object. Rather, it is the difference in gravitational force from one end of the object to the other. The gravitational field of a small low-mass black hole exerts fantastic tidal forces near its event horizon because the field declines in strength quickly as you move away from the hole. On the other hand, an enormous black hole extends powerful gravitational tendrils far into space. The gravity remains strong over long distances, and so the tides above the event horizon are not nearly as pronounced.

Consequently, some black holes are so bloated that they gulp entire stars without tearing them to atoms. Each star's gravitational potential energy converts entirely to speed—like the glass falling from the dinner table just before it hits the floor. None converts to heat and radiation. With no outpouring of energy, such black holes would be invisible. Calculations show that this intriguing shutoff valve may kick in when a black hole becomes about a billion times more massive than the Sun.

Under these circumstances, quasars are just early chapters in the lives of the cores of ordinary galaxies. As the central black hole ages and grows larger, the quasar becomes less hyperactive—and the galaxy around it looks more and more like a normal quiet galaxy. If that's true, supermassive black holes should be common, whether or not the cores of the galaxies are energetic. A growing set of data supports this notion: The list of nearby galaxies that appear to host dormant black holes has grown to more than two dozen. Our own Milky Way is among this fraternity. Astronomers believe that our galaxy's core probably contains a black hole a few million times more massive than our Sun. The giveaway is the vigorous speed that stars reach as they orbit near (but not too close to) the sleeping beast. Without such detailed observations, it would be hard to imagine that our deceptively peaceful galaxy harbors a heart of darkness.

Evidence for Supermassive Black Holes

Astronomers cannot use visible light to peer directly at the core of our own Milky Way or any other galaxy. The view is obscured by dust, gas, and countless stars that form seemingly impenetrable sheets of brightness. X-rays and radio waves, however, can pierce those veils. Images of the cosmos taken at those wavelengths have convinced astronomers that black holes as massive as millions or billions of Suns lurk at the cores of many—and maybe most—galaxies in the cosmos.

The celestial beacons known as quasars are several hundred billion times brighter than normal stars, and they shine across distances of billions of light-years. Many quasars reside at the cores of apparently normal galaxies, such as the quasar PG 0052+251 (*right*), about 1.4 billion light-years away. However, intense x-rays and radio waves from quasars suggest that the cores of these galaxies are far from normal. Of all the models devised to account for the prodigious energies of quasars, the most successful calls for supermassive black holes that swallow up to 10 Suns' worth of mass each year; radiation streams from blazingly hot disks of gas spiraling into the holes.

One of the strongest sources of radio waves in the sky is M87 (*opposite*), a huge galaxy best known for a narrow jet that blasts into space for hundreds of thousands of light-years. Measurements of this jet show that its particles move at nearly the speed of light, accelerated by a powerful cosmic engine at the galaxy's core. Spectroscopic observations with the Hubble Space Telescope zeroed in on the whirling motions of gas and stars near the center of M87. The gravitational forces needed to create such motions imply that a concentration of matter with 3 billion times the mass of our Sun sits within the galaxy's innermost 10 light-years. That virtually rules out any explanation other than a black hole.

Portraits of M87

The crown jewel of the Virgo cluster of galaxies is the prosaically named M87, a giant elliptical cluster about 50 million light-years away. A ground-based optical photograph (*below left*) shows M87 as a glowing ball of many hundreds of billions of stars. The Hubble Space Telescope peered more sharply into M87 to capture an image of the cosmic blowtorch that shoots in a narrow jet from the galaxy's core (*above right*). A radio portrait of M87's core (*below*), produced by an array of 27 radio telescopes, reveals intricate swirls and bubbles that emit torrents of radio waves. Astronomers believe that a supermassive black hole at the galaxy's center powers these dynamic features.

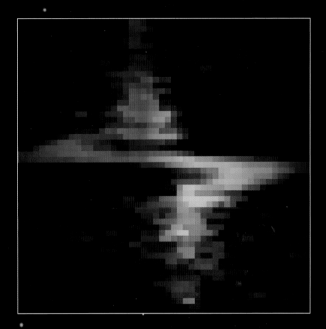

Bright Signatures of Dark Cores

An artist's conception of the hot gas falling into a supermassive black hole (*left*) shows how the whirling matter tends to form a flat disk, like the one that created our solar system. The thick gas prevents radiation from escaping out through the sides of the disk. Rather, the radiation blasts into space above and below the disk along narrow jets, propelling matter at close to the speed of light. When one of these jets from a distant quasar happens to point toward us, we call the intensely radiating galaxy a "blazar."

Astronomers can see the characteristic rotational signatures of such disks at the cores of many nearby galaxies. Shown above is a spectrum taken by the Hubble Space Telescope of the inner 26 light-years of the galaxy M84. The Doppler effect (page 139) reveals that stars and gas on one side of the galaxy's core are whirling quickly toward us at a rate of 250 miles per second (*purple pixels*). On the other side of the core, the stars and gas move away from us at the same rate (*red pixels*). The physics of gravity suggests that these rapid motions arise from the relentless pull of a central black hole at least 300 million times more massive than our Sun.

Frontiers

The Limits of Motion, Matter, and Energy

A steaming, red-tinged pool is not what most terrestrial life forms would call home. That creatures survive and thrive in hot springs like this one in Yellowstone National Park opened our eyes to the possibility of finding life where previously we would not have thought to look.

L

ate in the twentieth century, a small unmanned spacecraft flew past a rocky planet. Onboard sensors scanned the planet and found some unusual patterns. The atmosphere was rich in oxygen and methane, two highly reactive gases. Green light from the planet's star reflected into space off a substance covering most of the land. Organized radio signals streamed from the planet—transmissions that natural processes can't produce. The scientists who operated the spacecraft regarded all of these patterns as unmistakable evidence of life.

The planet, of course, was Earth. The spacecraft was the planetary probe *Galileo*, which spent the last half of the 1990s studying Jupiter and its dramatic moons. Mission scientists used Earth and Venus as gravitational slingshots to boost *Galileo* on its journey to Jupiter. During one of the Earth flybys, a team led by the astronomer Carl Sagan conducted a fascinating test: Could a spacecraft detect life on a planet without landing on it?

In Earth's case the answer was reassuring. The atmosphere provided two strong clues. Plants have altered the ingredients of Earth's air, making our atmosphere unique in the solar system. The most notable ingredient is oxygen, which composes 21 percent of the air we breathe. Free oxygen molecules react with rocks, soils, metals, and chemical compounds in the air. Rusty old cans in your garage are evidence of that process, called oxidation. Such reactions quickly remove most oxygen from the atmospheres of other planets. On Earth, however, another constant process churns the gas into the air just as quickly: photosynthesis. Earth's greenery has released enough oxygen to maintain a concentration of that vital gas in the atmosphere for the past 2 billion years.

Another unusual gas, methane, exists at much lower levels in the air—about 1 part per million. Methane doesn't last long in the presence of oxygen because oxygen molecules combine readily with methane to make water and carbon dioxide. Therefore, something on the ground must continually replenish the methane, just as plants do for oxygen. Volcanic eruptions and other natural events produce some methane but not enough to account for the levels seen. To fully explain the surplus, we must turn to biological sources: bacterial life in bogs (hence the name "marsh gas") and in the guts of termites, cattle, and other animals.

The light reflected from Earth's surface was a third clue. Large green patches detected on the continents by *Galileo* meant that some substance trapped other colors—primarily red and blue light—from the Sun. No known soils or minerals act in that way. The only logical explanation was widespread plant life, using chlorophyll to harness the Sun's energy. Pigments within algae, grasses, and leaves absorb mostly red and blue light to drive the chemical reactions of photosynthesis. The plants then reflect the remaining wavelengths of light, in the green part of the visible spectrum. That gives our planet its distinctive tinge.

A spacecraft flying past Earth 2 billion years ago would have seen these clues that life had taken root on the planet, Sagan's team noted. A fourth sign arose during the twentieth century, and it was the dead giveaway. An instrument on *Galileo* registered strong signals within narrow radio bands. The signals were too orderly to flow from the turbulent magnetosphere around Earth or from lightning or other natural bursts of energy. The radio waves also seemed to carry information: They pulsed and varied with distinct patterns. Those modulations, the researchers concluded, were hallmarks of the web of communication woven by an intelligent society.

Plant life on Earth absorbs mainly red and blue wavelengths of light from the Sun to drive the chemical reactions of photosynthesis. The remaining wavelengths, primarily green, reflect into space—giving our continents their distinctive hue and providing one strong clue about the presence of life on our planet.

We can use the observations of *Galileo* and our knowledge about the many forms of life on Earth to guide our quest for life on other worlds. If technological civilizations exist on planets around other stars, we may have a chance to detect their communications with radio telescopes on Earth. However, finding simpler forms of life elsewhere poses a challenge. We don't yet have the technology to measure the atmospheres of planets beyond our solar system, nor can we see the colors of light reflected from such planets. Within our solar system, no mission to other planets or moons has spotted any of the obvious signs of life that *Galileo* detected on Earth. But we know that life here has many other guises. Some organisms live far underground, drawing chemical energy from minerals and warm fluids. Others eke out bare existences within rocks in frigid Antarctica. Bacteria grow in the scalding chemical stews of hot springs in Yellowstone National Park and other geothermal sites. Most intriguing of all, ecosystems thrive on fiery ridges thousands of feet deep in the ocean, where new chunks of seafloor ooze

Life on the Edge

As we explore our own planet, we find that life has taken root in even the most extreme environments. Earth is home to organisms that thrive in boiling liquids and grow without light deep in the ocean, where they are also subject to intense pressures. Our planet hosts a tiny bug that can endure conditions nearly as severe as those found in space, microbes in Antarctic ice, and moss that remained alive yet dormant while frozen for 40,000 years in the Siberian permafrost. These organisms redefine the notion of life-supporting environments.

A Tiny Survivor

The pinhead-sized tardigrade (*left*), which lives in moss and mud in roof gutters and the cracks of paving stones, can withstand pressures 6,000 times greater than at sea level and temperatures from near absolute zero to 250 degrees Fahrenheit. It also survives complete dehydration as well as laboratory exposure to a vacuum and to intense x-rays. Some tardigrades have been revived after lying dormant in dried moss in museums for more than 100 years.

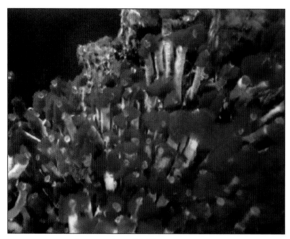

Fountain of Life

More than a mile beneath the surface of the ocean, thermal vents (*right*) offer an unexpected oasis for creatures such as crustaceans, bivalves, and giant tube worms (*above*). These organisms are part of a geochemical ecosystem depending on bacteria that use hydrogen sulfide from vent water as their primary energy source. The water, superheated under pressure to as much as 700 degrees Fahrenheit, dissolves minerals from hot magma welling up through cracks in the ocean floor.

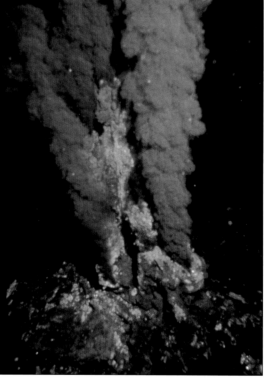

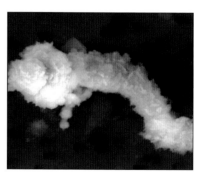

Cold Lovers

Colonies of organisms such as the cyanobacteria shown above exist beneath polar ice. Researchers liken these conditions to those on Mars. Dormant ancient microbes and even plants such as moss can remain preserved in ice, resuming metabolic activity after thousands to millions of years.

Heat Lovers

Organisms such as the bacterium *Moorella obsidium* (*left*) thrive in the near-boiling waters of geothermal hot springs in Yellowstone National Park. Gaining energy by oxidizing sulfur around the springs, such organisms produce sulfuric acid as waste.

out of the planet. Until recently, biologists didn't know that life could flourish in such extreme settings, often far beyond the reach of sunlight. These organisms—so-called extremophiles—have raised hopes that Earth's biological garden is just one of many oases in an otherwise barren cosmos.

Whether life has sprouted in other parts of the universe is one of the great cosmic questions we face today. We also seek to understand the behavior of matter on the tiniest scales and the origin and fate of the universe itself. Addressing these mysteries requires us to push our scientific skills to their limits. But if we continue to make steady progress in our explorations of motion, matter, and energy in the universe, even these answers may lie within reach.

Does Matter + Energy = LIFE?

At first glance the topic of life seems far removed from our discussion of the basic physics of the cosmos. However, all living things behave according to the laws of physics, no matter what or where they are. In 1917 the Scottish biologist D'Arcy Thompson noted in the introduction to his classic book, *On Growth and Form*: "Cell and tissue, shell and bone, leaf and flower, are so many portions of matter, and it is in obedience to the laws of physics that their particles have been moved, moulded and conformed." Thompson also recognized that those laws extend beyond the bounds of Earth: "Everywhere Nature works true to scale, and everything has its proper size accordingly. Men and trees, birds and fishes, stars and star-systems, have their appropriate dimensions . . . ." Thompson wasn't suggesting that life might exist elsewhere. Nevertheless, his thoughts provide a framework for our scientific inquiries into extraterrestrial life.

Thompson's observations cast a biological light upon two principles that guide our understanding of the physical nature of the universe. The Copernican principle, named for Nicolaus Copernicus, asserts that Earth is not the center of the cosmos. Rather, it is an ordinary place. Planets like Earth probably circle stars like our Sun throughout our galaxy and in similar galaxies elsewhere. We don't yet have the means to detect Earth-like planets, but we may within the next decade or two. On a more sweeping scale, the cosmological principle holds that the universe is the same everywhere, on average. The same laws of physics apply throughout space, holding sway over the interactions of matter and energy. For these reasons the processes that ultimately led to life on Earth could reasonably lead to life anywhere else in the universe, provided that the conditions are suitable to support life as we know it.

So what makes life precious and rare, limited only to Earth at least as far as we can tell? One answer simply may be that we have not yet looked long enough for life elsewhere or in the right places. If that's the case, the question may seem as quaint a few decades from now as the old belief that all heavenly bodies revolved around Earth. However, the answer may also lie in just how improbable it is for life to have arisen. Even the simplest organisms are incredibly advanced machines. To function and grow, an organism must maintain orderly processes within and around itself. Then, it reproduces and transfers that ability to its offspring. Such a chain of events runs strongly against a powerful tendency for systems to become more disorderly with time.

We know that tendency as one manifestation of a powerful principle called the second law of thermodynamics. According to this law, the disorder of a system—a quantity known as "entropy"—must increase with time if no energy enters or leaves the system. When entropy increases, things fall apart and lose their structure. For instance, a sugar cube dropped into a glass of water slowly dissolves. Eventually all the sugar molecules drift evenly throughout the water. We never see the reverse: a sugar cube assembling from floating molecules to appear miraculously at the bottom of the glass. At first glance the genesis of a living cell from a stew of organic molecules seems an equally miraculous violation of the second law of thermodynamics. However, the system is not isolated; it absorbs energy from the Sun and its environment. Many millions of years of mixing Earth's primitive organic ingredients with such inputs of energy led to the first cells. Billions of years of further interactions created the diverse life we see around us today.

Still, when we consider life in terms of entropy, it seems that the chances of creating a living biological machine just by following the laws of physics are ridiculously small. Yet life exists, so the probability is not zero. Does this mean life is plentiful? Will we someday make contact with other beings like ourselves? These questions often prompt spiritual responses, and understandably so. They cut to the core of our wonder about the universe and whether we are part of a grand cosmic design. Debating that metaphysical issue is easy; resolving it scientifically is impossible. However, we can and should use science in another way: to examine the sequence that matter and energy must follow to create a technological civilization out of formless interstellar gas. How plausible is each step along that path? The American astronomer Frank Drake posed that question mathematically in what is now known as the Drake equation. His formula offers a way to estimate the number of civilizations in our Milky Way galaxy that have the technology needed to talk to one another. The equation is a string of numbers and fractions. Put into words, it reads something like this:

Start with the number of stars that form in the galaxy each year. Multiply that rate by the fraction of stars with planets. Multiply by the number of planets or moons in each planetary system with conditions suitable for life. Multiply by the fraction of such planets upon which life has evolved. Multiply by the fraction of life-bearing planets with intelligent societies. Multiply by the fraction of those societies that have developed the technology to communicate across space. Finally, multiply by the average number of years that a technological civilization survives.

This long string of multiplications yields not a solid number but a range of numbers that depend on the assumptions you make at each step. Astronomers know some of the quantities fairly well, but they can only guess at others. For example, the galaxy gives birth to about 10 new stars each year. The growing number of planet discoveries seems to show that many stars, if not most, have planets. How frequently life arises on these planets is much less clear. Biologists believe that life took hold on Earth within a few hundred million years of the planet's birth, soon after a steady bombardment by large comets and meteorites died down. We don't yet know which chemical reactions spawned the first living cells. But chemistry is, at its roots, a special kind of physics that applies to atoms and molecules interacting with their environment in large groups. If physics is uniform throughout the universe—and if the Copernican principle is indeed a fundamental tenet of the cosmos—the chemistry of life may bubble forth in more places than many of us suspect.

Even so, planets experience a vast range of physical conditions. Which conditions are just right for life to arise? Here, we can draw lessons from the tale of Goldilocks. Her mischievous exploration of the bears' cabin led her to taste three bowls of porridge.

So what makes life precious and rare, limited only to Earth at least as far as we can tell? One answer simply may be that we have not yet looked long enough for life elsewhere or in the right places.

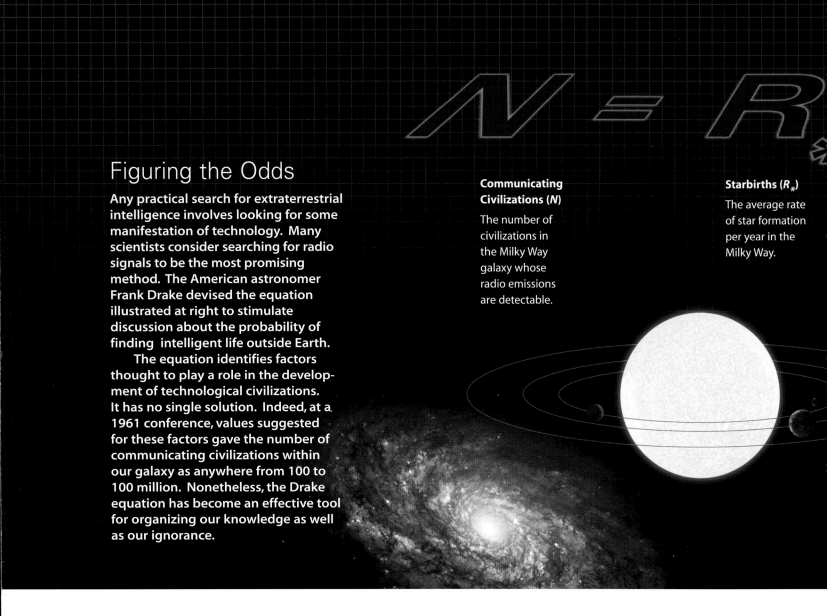

Figuring the Odds

Any practical search for extraterrestrial intelligence involves looking for some manifestation of technology. Many scientists consider searching for radio signals to be the most promising method. The American astronomer Frank Drake devised the equation illustrated at right to stimulate discussion about the probability of finding intelligent life outside Earth.

The equation identifies factors thought to play a role in the development of technological civilizations. It has no single solution. Indeed, at a 1961 conference, values suggested for these factors gave the number of communicating civilizations within our galaxy as anywhere from 100 to 100 million. Nonetheless, the Drake equation has become an effective tool for organizing our knowledge as well as our ignorance.

Communicating Civilizations (N)

The number of civilizations in the Milky Way galaxy whose radio emissions are detectable.

Starbirths (R_*)

The average rate of star formation per year in the Milky Way.

As you undoubtedly recall, one bowl of porridge was too hot, another was too cold, but the third was just right. Temperatures vary on the surfaces of planets as well. Thermometer readings on other worlds depend mostly on their distances from their parent stars and the compositions of their atmospheres. Planets very close to their stars fail the Goldilocks test, as do those orbiting in the deep freeze far away from their suns. For life as we know it, the "just right" range encompasses the temperatures at which liquid water can exist. At ordinary atmospheric pressures, that range is 32 to 212 degrees Fahrenheit.

This emphasis on liquid water is not arbitrary. Hydrogen and oxygen—the ingredients of the water molecule—are the first and third most common elements in the universe. Liquid water provides a medium within which chemical reactions can occur stably and quickly. That's not the case for the other forms of matter that water usually assumes, ice and vapor (or steam). The crucial chemical reactions involve substances that contain carbon, the fourth most common element. (The second most common, helium, is inert and plays no role in living things.) Carbon-based chemistry

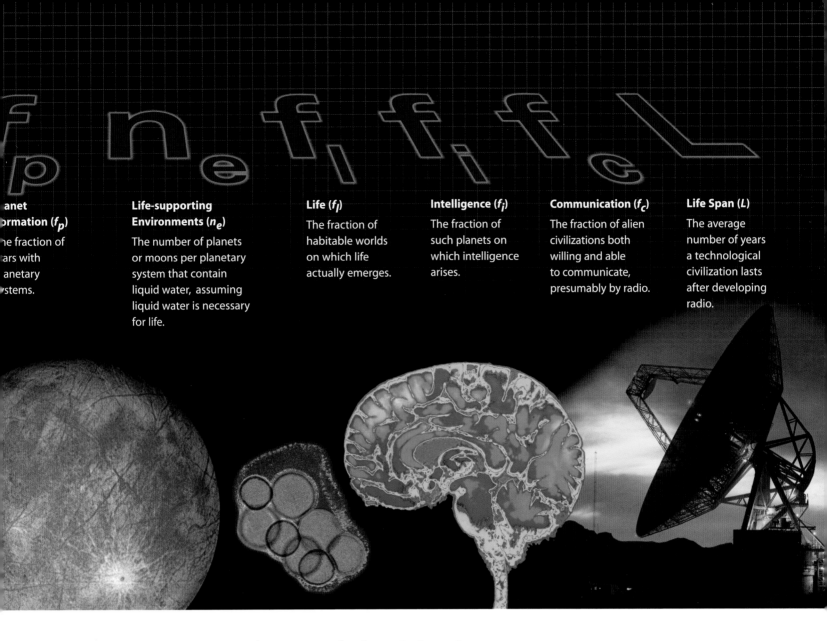

anet
ormation (f_p)

he fraction of
ars with
anetary
ystems.

Life-supporting Environments (n_e)

The number of planets or moons per planetary system that contain liquid water, assuming liquid water is necessary for life.

Life (f_l)

The fraction of habitable worlds on which life actually emerges.

Intelligence (f_i)

The fraction of such planets on which intelligence arises.

Communication (f_c)

The fraction of alien civilizations both willing and able to communicate, presumably by radio.

Life Span (L)

The average number of years a technological civilization lasts after developing radio.

is the most versatile chemistry in the universe. The element's chemical structure allows it to bond readily and strongly with itself and with many other elements. As a result, "organic" carbon-bearing compounds form the basis of all life with which we are familiar. It's encouraging that we see organic compounds everywhere we look. They exist in interstellar gas, dust grains, comets, meteorites, and on some moons in our solar system. Space is chock-full of the raw materials of life.

We must be careful not to rule out other possible building blocks of life. Perhaps silicon-based life covers a distant planet, powered by electricity rather than blood or sap. Indeed, our biggest challenge in the search for extraterrestrial life may be to shed our preconceived notions of what forms life must assume. But for now we know of no basic recipe for living things other than liquid water and carbon-based molecules.

With that in mind, one look around Earth is enough to see that we are safely within the Sun's "just right" range for life—a region that we more properly call the habitable zone. Rain falls, the oceans ebb and flow, and life teems within any

Redefining "Habitable"

Liquid water is still the prerequisite for any form of life that we have so far encountered. For a planet or moon to have sufficient heat to maintain liquid water, scientists once believed that it had to orbit within an optimal distance of its home star.

    Galileo now brings evidence that the habitable zone may not be defined solely by an orbiting body's distance from its sun. It has revealed that three of Jupiter's rocky inner moons—Europa, Ganymede, and Callisto—may harbor liquid water beneath their icy crusts. The tugging and pulling these moons endure in their gravitational dance with Jupiter and with one another may generate enough internal heat to keep water liquid despite the lack of solar warmth. And some scientists believe primitive life-forms may dwell at the base of Martian polar ice caps, where the planet's internal heat has melted the permafrost.

In 1976 the *Viking 1* lander showed early morning water frost or snow on Martian rocks. Mars may have regions of permafrost where water ice has been locked in the soil for millions of years.

A close-up of Europa's surface shows blue areas thought to be pure water ice. Darker mineral-laden water or slush from underground appears to have percolated to the surface through cracks in the crust.

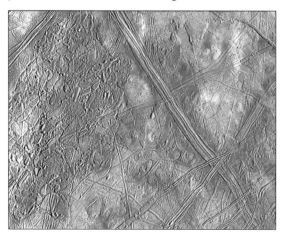

The surface of Callisto (*right*), which should be scarred by meteoritic craters, is instead relatively smooth and blanketed by something loose and fine grained. Upwelling ice may sublimate, leaving a residue of loose dirt, just as a snowman melts and leaves behind a little pile of soil.

puddle or tide pool. Some researchers think that Earth had an added advantage as a cradle for life: the Moon. Tides may have stirred the organic stew along Earth's early seashore, helping the first cells assemble and gain a foothold. Perhaps a large nearby satellite is another factor that nudges a planet toward habitability.

What about other places in our solar system? Venus also lies within the Sun's habitable zone. However, its out-of-control greenhouse effect has pushed temperatures too high (page 125). Mars appears cold and lifeless today. Yet its surface shows the dry remains of countless meandering riverbeds, deltas, and floodplains. There's little doubt that liquid water flowed on Mars in the distant past. We don't yet know how long that wet era lasted. Conceivably, life could have appeared on Mars before the planet became a vast Sahara. Earthlings were stunned in 1996 when researchers claimed that a potato-sized chunk of Mars called ALH 84001 contained fossilized evidence of primitive life. The rock struck Antarctica 13,000 years ago as a meteorite. It contains minerals that may have been altered chemically by microbes, plus tiny structures that resemble ultrasmall "bacteria." However, other scientists have disputed those claims. It now seems that the initial announcement about ancient life on Mars may have been premature. The final word on whether tiny extraterrestrials once colonized the Red Planet must await future missions there.

As we venture farther out into the solar system, it appears that all hope is lost for remaining in the Sun's habitable zone. Temperatures plunge to –150 degrees Fahrenheit or colder, and light dwindles to the intensity of moonlight. Yet we have discovered that liquid water can exist in such harsh settings. The most intriguing case is Europa, Jupiter's fourth-largest satellite. Tidal forces from Jupiter's intense gravitational pull flex Europa with relentless to-and-fro motions as the moon orbits the planet. Those motions create a potent source of internal heat. Images from the *Galileo* explorer suggest that giant slabs of ice cover an ocean of liquid water or possibly slush. This ocean may lie under at least 5 miles of ice, far too deep for sunlight to penetrate. But the energy from Jupiter's tidal forces may be sufficient to sustain life on the ocean floor, just as warmth from Earth's interior supports dark colonies of life along its volcanic midocean ridges. Astronomers envision future space missions in which robots drill through Europa's icy shield to search for such organisms.

To stretch the notion of a habitable zone to its extreme, consider what might happen if an Earth-like planet is ejected from its orbit around a star. Studies suggest that such ejections may be common. Jupiter-sized planets are gravitational bullies; they can fling their smaller brethren into deep space as casually as a flick of the wrist can snap a whip. According to calculations, those interstellar wanderers could retain their atmospheres. Thick atmospheres would trap warmth from the radioactive decay of elements and from volcanic activity on some of the planets. Those sources of heat could keep the oceans liquid, even without sunlight. The fascinating consequence is that life could persist on the seafloors of such nomadic worlds for billions of yers.

Clearly, there are many niches in the universe in which life could arise. This offers some hope that the first few fractions in the Drake equation—the ones involving the likelihood of planets and life—have reasonably high values. But what of intelligent life? That's a tougher question to tackle. Here, our methodical consideration of matter and energy doesn't help. Some researchers argue that, given enough time, the progression from primitive life forms to intelligence is inevitable. Others observe that billions of species of life in Earth's history have led to just one society capable of beaming signals across space. And then there is the last factor in the Drake equation: the average lifetime of such a civilization. If other technological societies last only a century or so before self-destructing—as we nearly have upon occasion—the galaxy will contain few of them at any one time. But if they manage to survive for millions of years or longer, our Milky Way could be a haven for advanced aliens.

If any of these societies wanted to communicate, we presume that radio waves would be their band of choice in the electromagnetic spectrum. Radio signals are easy to create, and large radio telescope dishes can beam them on tight searchlight cones into space. Most important, radio waves traverse the galaxy unimpeded by interstellar gas and dust. Several groups of astronomers are systematically training their radio telescopes on nearby stars to try to pick up such signals. Their projects are collectively called the Search for Extraterrestrial Intelligence, or SETI. The most advanced efforts use powerful electronic detectors that can monitor billions of narrow radio channels at once, sifting for signals that rise above the cosmic din.

It seems that we Earthlings are primarily in "receive" mode when it comes to SETI. We rely upon other civilizations to beam greetings (or some other message) to us. But if

Imprints of Life?

Meteoritic visitors not only gouge huge craters in Earth's surface, they also carry with them material from elsewhere in the solar system. For example, a piece of Mars (*below right*), has been carefully preserved and thin slices apportioned to researchers who hope to learn whether tubelike structures seen under the microscope may be fossils of primitive organisms. A meteor that landed in 1998 in West Texas was found to contain blue and purple halite (*right*), a crystal similar to table salt. Further analysis confirmed the presence of water inside the crystals. Radiation has darkened the crystals, which are estimated to be 4.5 billion years old—meaning that the briny water could predate the Sun and planets in our solar system.

Other clues to chemicals in the early solar system could come from a complex class of organic ring molecules called polycyclic aromatic hydrocarbons (*below*). Spectral analysis has revealed the presence of these compounds in interstellar space.

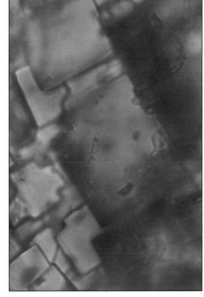

At less than a tenth of an inch across, the briny crystals in the West Texas meteorite are the largest halite crystals ever seen in any extraterrestrial material.

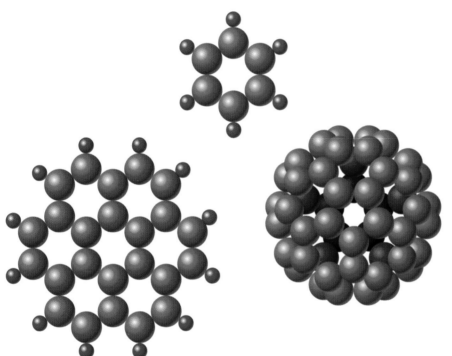

Labeled ALH 84001, this 4.5-billion-year-old rock was once part of Mars. It is a piece of a meteorite that landed in Antarctica 13,000 years ago after having been dislodged from Mars by a high-energy impact about 16 million years ago. Although the rock's dryness does not strengthen the case for a water-rich crust on Mars, scientists hope to determine whether it contains fossil evidence of primitive life that may have once existed on the Red Planet.

The structural basis of polycyclic aromatic hydrocarbons is the highly stable benzene ring, a hexagonal arrangement of six carbon atoms, each attached to a hydrogen atom (*top*). Cororene (*above left*) is one such molecule whose presence has been inferred from spectral signatures. Though not yet formally identified in space, an even more stable configuration, the fullerene (*above right*) may have been the first object with a surface to form after the Big Bang. Fullerenes would be strong enough to survive impact with other molecules and go on to build dust, comets, asteroids, and planets. Some types of hydrocarbons found in ALH 84001 (*right*) are considered candidates for prebiological organic molecules.

everyone is only set to receive signals and no one is beaming, would SETI be doomed to fail? Not necessarily. Even as we speak, our daily weather reports and reruns of *I Love Lucy* are streaming into space at the speed of light. Our radio and television signals fill a growing bubble of our galaxy, now extending more than 50 light-years from Earth— far enough to encompass hundreds of other stars. Although these broadcasts are much weaker than an intentionally beamed message would be, they might register in sensitive telescopes on the planet Grok, and vice versa.

Our species has taken some baby steps to communicate with our possible neighbors. In 1974 Drake and his colleagues used the giant Arecibo radio telescope in Puerto Rico to beam a three-minute message toward a cluster of stars. The message, encoded as a binary sequence, contained basic information about our planet, our solar system, and the biology of humans. Four of our planetary explorers—*Pioneer 10, Pioneer 11, Voyager 1*, and *Voyager 2*—carry greetings as well. Each *Pioneer* has an engraved plaque about us, while each *Voyager* features a gold-plated phonograph record with sounds and audio-encoded images from Earth. But these efforts are symbolic, not practical. For instance, the Arecibo signal will reach its target in 24,000 years. Even if a civilization there sent a reply, we wouldn't get it until the year 50,000.

Another approach to SETI relies not on radio waves but on visible and infrared light. Think of spotting the light from a handheld laser pointer across a football stadium. Even with all the light entering your eyes from the stadium lights, the playing field, and other sources, you'd probably notice the laser pointer right away. All its energy is concentrated into the familiar red laser color, and its beam remains intense as it crosses the field. In a similar way, powerful single-color lasers directed into space by a large telescope would be visible across great stretches of the galaxy. Brief but energetic pulses of laser light could outshine the parent star within that part of the color spectrum. Some SETI researchers have adapted their equipment to watch for such pulses.

It's fascinating to ponder the possible outcomes of SETI. If the project succeeds, our awareness of the universe will undergo a far greater upheaval than Copernicus caused by displacing Earth from the center of all things. If SETI fails after many years, our descendants would confront this challenging question: Despite the rich interactions between matter and energy throughout the cosmos, are we indeed alone?

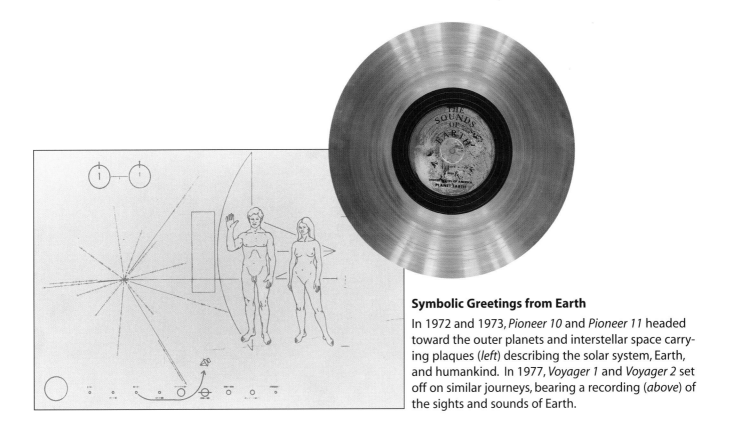

Symbolic Greetings from Earth

In 1972 and 1973, *Pioneer 10* and *Pioneer 11* headed toward the outer planets and interstellar space carrying plaques (*left*) describing the solar system, Earth, and humankind. In 1977, *Voyager 1* and *Voyager 2* set off on similar journeys, bearing a recording (*above*) of the sights and sounds of Earth.

Where Did the UNIVERSE Come From?

Long before humans wondered whether other beings lived among the stars, they questioned the origin of the universe itself. Cultures invented creation myths and passed them down through generations, forming the richest of all stories for anthropologists to decipher. Modern Western culture has devised a story of universal creation as well. This particular story goes beyond myth because we can back it up with scientific data.

The modern creation story is called the Big Bang. It is grounded in Edwin Hubble's discovery of the expanding universe in 1929. The American physicists Ralph Alpher and George Gamow were the first to propose the theory seriously, in 1948. According to Alpher and Gamow, the universe began with a burst of nuclear fusion from which all elements arose. Astrophysicists later learned that giant stars forge most elements heavier than helium, but the nugget of the idea was in place. It appeared the universe had an explosive birth and is still flying apart after billions of years, like a fireworks blast that literally fills all of space.

For a time, proponents of the Big Bang waged an intellectual war with supporters of another theory of the universe, called the steady-state theory. In this view, the universe obeyed the "perfect cosmological principle"—it had always and would always look the same everywhere, with no beginning and no end. As galaxies drifted apart from each other, new matter slowly arose out of empty space to keep the overall density of the universe the same. The steady-state theory was pleasing to the mind, avoided the question of a single "origin," and was consistent with observations of the universe for many years. Then, in the 1960s, astronomers discovered quasars. These bright beacons of light were all far away from Earth; none were nearby. Thus, they were more common when the universe was younger. This violated the perfect cosmological principle because it indicated that the universe had clearly changed with time—more evidence for the Big Bang.

Since then, data from a series of studies have continually supported the Big Bang. These studies have forced scientists and nonscientists alike to grapple with the theory's implications. Let's face it: The Big Bang is bizarre. It suggests that an explosion roughly 13 billion years ago created all space and matter and energy within a fireball that initially could have passed through the eye of a needle. By taking a hard look at the details, however, we will see why astrophysicists are willing to put their stock in the Big Bang.

Evidence in favor of the Big Bang rests on all three pillars of our approach to understanding our universe: motion, matter, and energy. Hubble's research revealed that distant galaxies recede from us more quickly than closer ones in direct proportion to how far away they are. We now recognize this as the expansion of space itself, launched by the Big Bang. Einstein's general theory of relativity predated Hubble's work by 13 years. Even so, solutions to one of the theory's many equations predicted a universe that expands precisely according to the pattern found by Hubble. Today, as we peer more deeply into space, we still observe that galaxies appear to flee from us more quickly as the distances grow larger—just as Hubble would have predicted.

How do we know? One clue comes from gravitational lenses. The gravitational field of a massive object can noticeably bend light. Sir Arthur Eddington proved that in 1919 by observing the warped paths of starlight passing close to the Sun during a total eclipse. If light from a distant quasar travels near a galaxy or a cluster of galaxies on its way to Earth, the gravitational lens can create three or more images of the quasar. The resulting optical effect is like looking at a warped mirror and seeing several images of your face. They're all you, but the light rays have traveled different paths to get to your eyes. We have observed such optical antics for dozens of quasars. Red shifts in the spectral lines of the objects reveal how quickly they are moving away from us. In each case the more distant "lensed" object is always traveling faster than the object whose gravity serves as the lens. We never see gravitational lenses in which the distant quasars move more slowly than the closer lensing galaxies. In other words, gravitational lenses support Hubble's contention that the expansion of the universe grows faster and faster with increasing distance.

Einstein's special theory of relativity provides another clue. Think back to the hyperkinetic unicyclist pedaling past you at close to the speed of light. His mass increased, his length shrank, and his clock slowed down relative to yours. We have identified "clocks" in distant galaxies that display this effect in space. The clocks are supernova explosions, which behave uniformly from one galaxy to the next. The most distant supernovas take more time to explode and to decline in brightness than comparable ones in nearby galaxies. That's just what should happen if the distant galaxies stream away at a fair fraction of the speed of light. A "week" for us might appear to last eight days, nine days, or longer for supernovas in such galaxies.

These objects in motion all trace backward to a time when the universe was much smaller than it is today. The matter within these objects—the stuff of stars and galaxies—holds another important test. The Big Bang theory maintains that the fires of the early universe raged at trillions of degrees, too hot for atoms to exist. Instead, matter consisted of a soup of quarks, electrons, and other subatomic particles. It took about 100 seconds for temperatures to drop to a billion degrees, "cool" enough for the first atomic nuclei to fuse. We can take an educated guess at calculating the mixture of atoms that resulted. To do so, we combine all that we know about quantum mechanics with all we have learned about smashing atoms to smithereens in particle accelerators. The outcome predicts a universe with an original mix of about 75 percent hydrogen; 25 percent helium; and a smattering of other ingredients such as lithium, the third element in the periodic table, and deuterium, a type of heavy hydrogen with an extra neutron. This matches what we see in the universe to a satisfying degree.

The most convincing evidence for the Big Bang comes not from motion or matter but from the energy of the universe itself. If an unimaginably hot explosion formed the universe, it should have cooled ever since. Like the cooling cinders of a fire that once blazed red hot, space itself should radiate with some leftover warmth from the Big Bang. Calculations predict that this glow should permeate the universe at just a few degrees above absolute zero. In 1965 the physicists Arno Penzias and Robert Wilson found this residual heat serendipitously, the cosmic microwave background radiation.

A satellite called COBE—the *Cosmic Background Explorer*—confirmed this finding in 1992. The satellite measured microwaves in nearly all parts of the sky and found a background temperature of 4.9 degrees Fahrenheit above absolute zero. This was firm evidence that the universe has cooled uniformly since the Big Bang. The pattern of radiation was strikingly smooth; it varied by less than 0.0002 degree from one part of the sky to the next. These subtle variations in temperature are like tiny bumps, no more prominent than large grains of sand under a sheet the size of a football field. We interpret these ripples as the signatures of quantum-scale fluctuations during the Big Bang. As the universe expanded and cooled, the patterns served as seeds for galaxies and clusters of galaxies to form.

Finally, we can turn to something as simple as a temperature gauge to provide one more piece of evidence supporting the Big Bang. Distant galaxies—which we see not as they are, but as they were billions of years ago—should be bathed in a hotter cosmic microwave background than today's galaxies. To test this notion, we can use large telescopes to analyze the light from those faraway galaxies. Spectra of these galaxies are crisscrossed by many sets of lines, which reveal the composition of the gases and stars in the galaxies. The lines of certain hydrogen molecules are special: They reveal how quickly the molecules vibrate. The rates of vibration pinpoint the temperature of the environment there, just as reliably as the thermometer in your window. Sure enough, spectra of distant galaxies show that the microwave background in that long-ago era was several degrees warmer. The measurements match the gradual rate of universal cooling predicted by the Big Bang theory.

This array of evidence is impressive. The acceptance of the Big Bang represents an unprecedented unification of astrophysics and particle physics. A coherent cosmic picture, our modern creation story, has emerged from a minimum of assumptions and measurements. Still, most people react to the Big Bang by objecting that it doesn't make sense. How could the entire universe start as an unimaginably energetic speck?

It seems a powerful objection, but we must acknowledge that our ability to measure nature long ago outstripped our senses. No longer can we invoke common sense to evaluate whether something is true. For this reason, the fact that the Big Bang is so bizarre should not affect our willingness to embrace it. Quantum mechanics, another highly successful theory, was similarly received when it was first advanced. Quantum mechanics describes a world in which particles act like waves, obeying an odd set of rules that prevents certain motions or states of energy from existing. On these tiniest scales of all, our universe is nonsensical. It should be no surprise that the Big Bang, an explanation for all that exists on the largest scales, is just as strange.

Even so, many questions remain about the Big Bang. For example, how did the energy of the universe expand so smoothly? The tiny lumps seen in the cosmic microwave background are smaller than one would expect from an explosion as fierce as the Big Bang. The American astrophysicist Alan Guth proposed a solution in the early 1980s: "inflation." This scheme has no everyday analog, so don't confuse yourself by thinking of the rising costs of food or fuel. Cosmological inflation refers to an interval when the baby universe expanded at a wild rate. This period of time was incredibly brief—less than 1 billion trillion trillionth of a second. During that cosmic eyeblink, the size of the universe increased by an extraordinary factor of 10^{50}. When this furious growth stopped, the cosmos was about the size of a beach ball. Its expansion then "slowed" to its normal explosive pace. The inflationary process would have smoothed out the universe to the extent we see today. The hypothesis also makes specific predictions about the details of the cosmic microwave background and the patterns of gravitational collapse that caused the first groups of galaxies to form. Modern satellites, telescopes aboard high-altitude balloons, and other instruments will map variations in the microwaves with exquisite accuracy to provide a stern test of the inflation hypothesis.

Beyond such scientific details, several basic questions about the Big Bang may seem reasonable to ask. When we ponder the questions carefully, however, we realize that some of them don't make physical sense. For example, what lies beyond the universe? Or, put another way, what is it expanding into? We cannot answer that question because the universe contains all space. Space expands everywhere at once, and there is no "outside." Light beams are forever limited to the confines of our universe. Because nothing can travel faster than light, we cannot probe "beyond." In this way the entire cosmos is like the largest black hole of all, trapping all that exists within its

The acceptance of the Big Bang represents an unprecedented unification of astrophysics and particle physics.

vast event horizon. Here's another favorite topic: What came before the Big Bang? To answer that the laws of physics did not work before the Big Bang may sound as if we're dodging the issue. However, time and space were meaningless before the Big Bang; they simply did not exist. Just as you cannot go any farther north on Earth if you are standing at the North Pole, you cannot go farther back in time from the birth of the universe.

This does not stop some theorists from speculating. The cosmic microwave background tells us that the space-time fabric of our infant universe was roiled by fluctuations on the smallest possible scale—a quantum foam, if you will. One class of inflationary hypotheses describes a mega-universe with endless bubbles of expansion arising from this foam. Each bubble looks like a Big Bang universe from

Signs of the Big Bang

The cosmic microwave background is the faint glow of radiation that remains from the primordial fire-ball—the Big Bang—that gave birth to the universe. Within a fraction of a second, the ultimate character-istics of our universe were set. Space-time expanded, the energy of the universe spread thinner and thin-ner, and the average temperature dropped—swiftly at first, more slowly later on.

Big Bang Plus 10^{-43} Second

This artist's conception represents the cosmos in the tiniest fraction of a second after the Big Bang. At this time the universe was much smaller and hotter and contained minute irregularities—bumps no larger than grains of sand would make under a bedsheet the size of a football field. Most theories then call for a period of rapid expansion known as inflation.

within, and it may sustain laws of physics that differ from the ones we know. Inhabitants of those other bubbles would face the same impossible mysteries, forever confined to their part of the "multiverse" (page 178). Testing this wild scenario will require some scientific advance that we can't envision today. At this point it's safe to say that our laws of physics don't exclude the existence of other universes; they simply can't explain them.

It's tempting to regard the Big Bang as little more than science fiction. However, our belief in the theory is bolstered by an impressive set of successful predictions—far more than most theories in progress enjoy. Indeed, nearly everyone in the community of astrophysicists now accepts the Big Bang. But we all recognize that, as cosmology progresses, it one day may become the core idea of something even bigger.

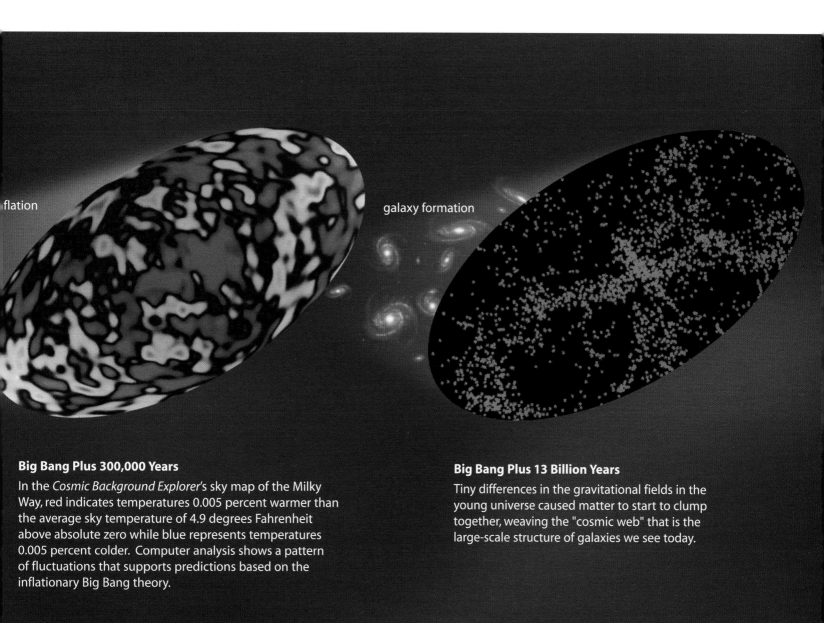

flation

galaxy formation

Big Bang Plus 300,000 Years

In the *Cosmic Background Explorer*'s sky map of the Milky Way, red indicates temperatures 0.005 percent warmer than the average sky temperature of 4.9 degrees Fahrenheit above absolute zero while blue represents temperatures 0.005 percent colder. Computer analysis shows a pattern of fluctuations that supports predictions based on the inflationary Big Bang theory.

Big Bang Plus 13 Billion Years

Tiny differences in the gravitational fields in the young universe caused matter to start to clump together, weaving the "cosmic web" that is the large-scale structure of galaxies we see today.

Multiple Universes?

One of the most intriguing questions in cosmology is: What came before the Big Bang? One hypothesis, illustrated here, is that our universe may be just one of many that materialized out of the inherent instability of the cosmic vacuum. These vacuum fluctuations may be similar to an era of fluctuating foam known as Planck Time, 10^{-43} second after the Big Bang, before which our current models and theories do not apply. Some researchers speculate that the writhing foam would have produced tiny bubbles that appeared and disappeared—or suddenly expanded into entire universes.

How **SMALL** Does Matter Get?

For another cosmic mystery, it's not necessary to peer billions of light-years into space or back to the beginning of time. This mystery is right at your fingertips: What is the essential nature of matter? From the atoms of Democritus to the atomic nuclei of Ernest Rutherford to the quarks of Murray Gell-Mann, our models of matter have shrunk to ever-smaller scales. It's natural to wonder whether this progression will continue, surprising us with even tinier nesting Russian dolls of matter. The answer is more than just a curiosity, because quantum mechanics tells us that energy and matter fluctuate on these tiniest scales. In the earliest moments of the Big Bang, continual quantum fluctuations dictated the future appearance of the universe—and they form the basis of all space and matter today.

Physicists peel back matter's layers by boosting electrons, protons, and other particles to exceptionally high speeds and smashing them together. Our Earthly machines are pale versions of natural particle accelerators in the cosmos. One of the closest is the Sun, which pierces the inner solar system with writhing magnetic fields. Charged particles in the solar wind zoom outward along these field lines like surfers catching steep waves. Pulsars, black holes in the centers of galaxies, and other energetic objects also fling particles into space at close to the speed of light. Accelerator energies on Earth are low by comparison, but they still allow physicists to simulate the conditions that existed fractions of a second after the Big Bang, albeit within volumes smaller than that of an atomic nucleus.

Decades of such research have constructed a scheme for the universe with a rather dry name: the standard model. This thorough model describes all known particles in the universe and their interactions with precision. Its main ingredients are the four forces of nature—gravity, electromagnetism, and the strong and weak nuclear forces—and two distinct sets of particles. One set consists of the basic units of matter as we know them today: quarks, electrons, and neutrinos. Each of these types of matter falls into one of three "families," divided according to their masses. For the most part, the stuff of our everyday world features particles in the least-massive family. The other two families arise mainly in particle accelerators, both on Earth and in space. The second set of particles in the standard model is fundamentally different.

It consists of particles that carry the four forces among the basic units of matter. Like waiters flitting from kitchen to table in an excellent restaurant, the force-carrying particles do their jobs without our really noticing them. We know about photons, which ferry electromagnetic forces from place to place. The other force-carrying particles stay mostly hidden from view.

The various building blocks of matter combine to create hundreds of composite particles. You've heard of two of these—protons and neutrons. It's less likely that you're aware of pions, kaons, and omegas. Not to worry; the standard model has everything under control. It tells us how all particles arise and how they interact, just as surely as we can glean the properties of a chemical element by glancing at the periodic table. Physicists are as comfortable with the standard model as chemists are with their table of neat rows and columns.

However, there's a major difference between the periodic table and the standard model. The quantum mechanical behavior of electrons within atoms tells us why the periodic table looks the way it does. Its patterns are the direct results of how electrons act when atoms approach one another. The standard model, on the other hand, lacks such a tangible foundation. Instead, it prompts questions that physicists cannot yet answer. For example, why are there three families of matter? Why aren't there two or four? What dictates the mass of each particle? A proton's mass is

Reconstructing
the Early Universe

In mammoth particle accelerators, beams of particles such as protons and antiprotons collide at nearly the speed of light, bursting into showers of quarks, electrons, neutrinos, and other subatomic particles. These experiments probe the nature of matter and energy and aid in the quest to unify the fundamental forces of nature.

A circular course 1.3 miles across marks the enormous underground Tevatron accelerator at Fermi National Accelerator Laboratory near Chicago.

1.67×10^{-27} kilograms—that's 270 trillion trillion protons per pound. An electron is about 1,835 times less massive. The standard model is full of numbers like these, but no one knows exactly why the values are what they are.

One major answer may come in the form of a hypothetical particle with a slightly comical name: the Higgs boson. This force-carrying particle would create a field that permeates the universe, much like magnetic or gravitational fields. Flecks of matter would acquire their masses by experiencing this field to different degrees. Imagine wading through three pools filled respectively with air, water, and molasses. You'd "feel" light, then heavier, then heavier still as the substances dragged on you. The Higgs field works in a similar way. Fleeting neutrinos would interact with the field not at all or just barely, electrons a bit more, and quarks considerably more strongly.

Physicists are racing to find the Higgs boson within the blasts of particles created in their colliders. If they detect it, the standard model will become an ever more powerful tool for understanding how matter assumes its many forms in the universe.

Despite the successes of the standard model, nearly all physicists agree that it will be subsumed by a more fundamental theory of how the cosmos ticks. Rules of physics beyond the model's tight confines should illuminate other enduring mysteries. One of these is why matter exists at all. The Big Bang theory predicts that the fury of the

Engineers fine-tune the Brookhaven National Laboratory's Relativistic Heavy Ion Collider in New York. Two rings of powerful superconducting magnets guide particles in opposite directions, culminating in collisions that yield new particles and states of matter. In one experiment, colliding gold nuclei momentarily create densities equivalent to that of the Earth if it were squeezed into a 30-foot cube. In the process, researchers hope to observe the creation of the quark–gluon plasma that may have existed moments after the Big Bang, long before matter formed into individual atoms.

explosion created equal amounts of matter and antimatter, particles with identical masses and spins as their matter counterparts but with opposite charges. But when we look around us today, we see precious little antimatter. That's a good thing, because a blob of antimatter would annihilate you in a burst of gamma rays if you touched it. Soon after the Big Bang some unknown interaction tipped the scales in favor of matter. The imbalance was tiny: For every billion particles of antimatter, there were a billion and one particles of matter. All of the matter–antimatter pairs were annihilated in an orgy of photons, leaving behind a smidgen of matter. Well, not quite. That "smidgen" comprises everything in the cosmos today. We owe our existence to the mysterious interaction that skewed nature toward matter. Physicists are in hot pursuit of this interaction as a possible first glimpse past the standard model.

Beyond that prize, the holy grail in particle physics today is the immodestly named "theory of everything." Such a theory would unite the forces of nature into a grand description of how all particles interact and why they have the properties they do. Normally, the four fundamental forces act separately from one another and with different strengths. But this is not always true. Physicists have shown that electro-magnetism and the weak force are different manifestations of a single force, called "electroweak." This force exists only when the energies of particles—which we can think of as their temperatures—get very high. How high? How about a quadrillion degrees, the temperature of the universe less than a billionth of a second after the Big Bang. Seeing this one force act as two forces in our everyday world is like seeing water exist as ice, liquid, vapor, or scalding steam, depending on the conditions. We recognize each substance as water, even though superficially they are not any-thing alike.

At even higher energies the other two fundamental forces may also merge. The strong force would join next but only at the temperatures that raged in the universe when it was just slightly more than a trillion trillion trillionth of a second old. That's way beyond the reach of our particle accelerators. Even so, physicists have developed descriptions of how the strong force, the weak force, and electromagnetism behave as

Deep in a tank of liquid hydrogen, a proton struck by a speeding neutrino disintegrates into a spray of subatomic particles in this artist-enhanced image from the European Centre for Nuclear Research (CERN) in Geneva, Switzerland. The reaction may approximate conditions that existed in the early universe.

one unified force in such extreme conditions. Such theories are called grand unified theories, or "GUTs."

A successful GUT still would fall short of a theory of everything because of one missing element: gravity. Gravity is the toughest nut to crack in theoretical models of the behaviors of matter and forces. It is so much weaker than the other three forces that it has little effect on the scale of atoms. And yet it is so far reaching that it determines the appearance and fate of the universe. In its current form the standard model cannot account for gravity at all. Unifying the cosmic realm of gravity with the microscopic realm of quantum mechanics would rank as the crowning achievement of modern physics. It's a quest that may take decades.

Today's best candidate for such a unifying description is called string theory. For a moment, suspend all preconceptions you have about matter to entertain what string theorists claim. In the standard model, we can think of particles as points of mass. *No*, say string theorists; particles actually are minuscule strings or membranes that vibrate in space. Each particle would represent a different mode of vibration of the strings, much as a single guitar string can create many notes. The forces of nature would arise from the harmonies of the interacting strings.

String theorists did not invent these notions for their symphonic analogies, although they are pleasing. Rather, the mathematical details of these notions come closest to the theory of everything that physicists so fondly envision. In particular, string theory requires gravity to exist—whereas the standard model says nothing about the origins of gravity. The problem is that we have never seen one of these strings, nor can we ever hope to. By all indications, they are at least a billion billion times smaller than the dimensions of a proton.

Not all is hopeless, however. Even stranger consequences of string theory offer some chance of testing it in our lifetimes. First, the theory predicts that all particles have symmetric partners. These partners are not the same as matter and antimatter. Rather, every particle of matter would have a symmetric force-carrying counterpart, and vice versa. Given their penchant for superlative prefixes, theorists have named this notion supersymmetry. No such shadow partner has yet emerged from our particle accelerators, but the search goes on.

A Universe of Strings?

According to string theory, if we could look more deeply into fundamental particles such as electrons and quarks—something currently well beyond the abilities of the most powerful particle accelerators—we would discover that these particles are actually one-dimensional loops, or strings.

All matter is made up of atoms. Atoms, in turn, are made up of electrons orbiting a nucleus of protons and neutrons, which themselves are made up of quarks. String theory suggests that electrons and quarks are actually minuscule vibrating loops of energy.

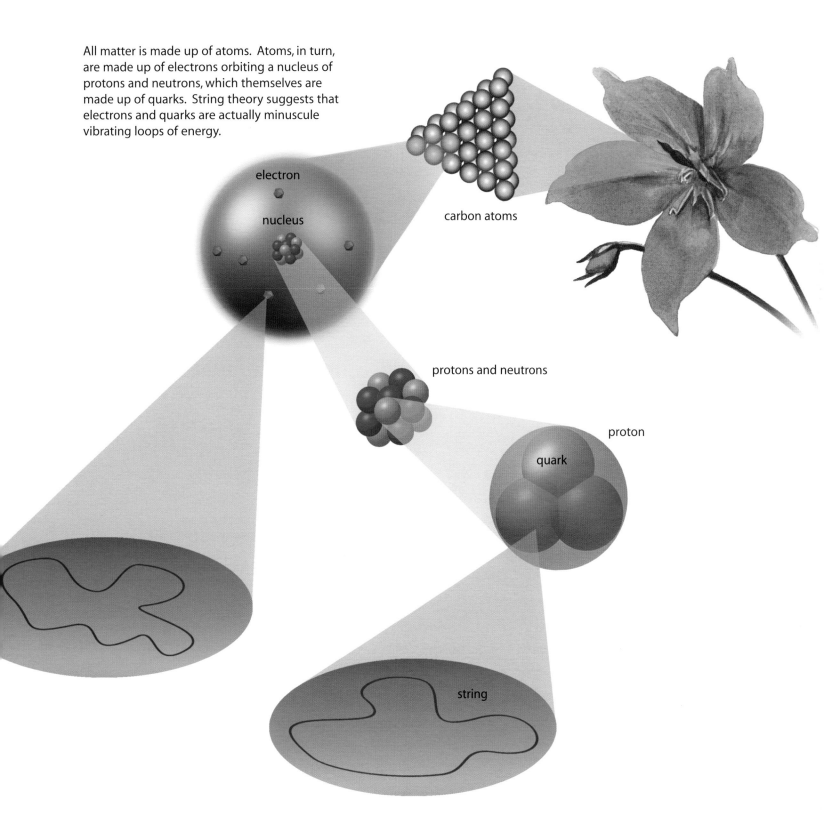

electron

nucleus

carbon atoms

protons and neutrons

proton

quark

string

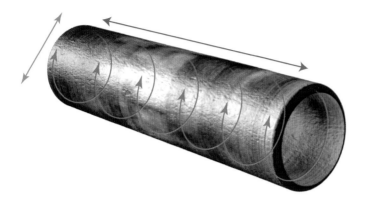

Hidden Dimensions

The mind-stretching concept of more than three spatial dimensions is actually demanded by the mathematics of string theory. As illustrated here, these additional very small spatial dimensions may be contained within the familiar dimensions we commonly experience. If the drainage pipe at left were seen from a great distance away, it would look like a one-dimensional line. This horizontal dimension is emphasized by the blue arrow. At closer range a second dimension, width, becomes apparent (*green arrow*). Finally, as we approach even closer, we perceive the pipe's depth (*curving arrows*). We might say that the second and third dimensions were "curled up" inside the pipe. In similar fashion, additional dimensions might become visible if we could turn up the magnification on the fabric of space-time (*below*).

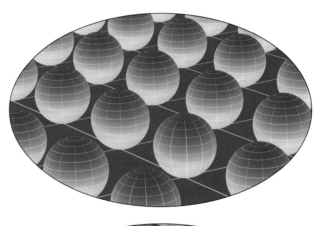

Each level at left represents an increasing magnification of the fabric of space-time. The grid lines represent the dimensions we ordinarily experience. To convey the idea of where the extra dimensions of string theory might hide, the top level shows how one additional dimension might be viewed as circles, akin to the loops in a very thick pile carpet. For simplicity the circles are shown spread out at the intersections of the grid lines, but in reality they would exist at every point. Two additional curled-up dimensions are represented above as a sphere and a torus, or doughnut.

Second, string theory predicts that six or more extra dimensions exist beyond the four dimensions of space and time with which we are familiar. If you have a hard time picturing the four-dimensional weave of space-time, you may throw up your hands at the concept of six more dimensions. But as a real-world analogy, imagine driving on a flat highway and seeing a large drainage pipe jutting out into a field next to the road. From far away the pipe looks like a one-dimensional line emerging from beneath the asphalt. As you approach, the pipe gains width; it appears two dimensional. Finally, if you stop your car and walk into the field, you see the pipe's complete shape. You can even climb inside it, a three-dimensional experience that only opens up when you touch the pipe instead of looking at it from afar. In a similar way, the six extra dimensions of space are curled up—"compactified" in the lexicon of physics—under ordinary conditions. We can't see them unless we get up close. Once again it's an issue of energy. Some physicists think the accelerators of today or the near future will open up these hidden dimensions. New particles might pop out, and gravity might behave very differently from the way it does on large scales. If we see such effects, we'll be one step closer to writing our grand recipe for the universe.

The Source of Big EXPLOSIONS

If extra dimensions do exist, they would have contributed to the fabric of space during the earliest moments of the Big Bang. But no matter how far out into the universe we look—that is, no matter how far back in time our telescopes can peer—we will never see those first incandescent moments directly. The cosmic microwave background blocks our view. The microwaves, which provide us with such solid evidence of a once-hot universe, date to a time when the universe was roughly 300,000 years old. At that point the entire cosmos was about as hot as the surface of our Sun, about 10,000 degrees Fahrenheit. The universe was opaque; light scattered back and forth within the plasma, just as photons ricochet within the Sun's body. When the temperature dropped below that level, the universe became transparent to light because the first hydrogen atoms formed, allowing photons to stream through without striking loose electrons. The first photons that traveled freely away from the hot smog make up the remnant microwaves that our telescopes see today.

In other words, the microwave background is a photon screen that stops us from looking back in time all the way to the Big Bang itself. Clever scientists of the future may devise ways to penetrate that veil. They may learn to detect neutrinos that flashed into space when the first elements fused or gravitational ripples in space-time that were created by the Big Bang. Fortunately, we don't need to wait for new technological advances to study gigantic explosions. Other blasts, quite visible to us today, are the biggest bursts of energy in the cosmos since the primeval Big Bang. We call them gamma-ray bursts.

How bright are gamma-ray bursts? If we floated above Earth's atmosphere and had gamma-ray eyes, we would see flares of light as bright as Venus blaze in the heavens once a day. After mere seconds they would vanish forever. They are so far away that to appear as bright as they do their energy releases must be enormous. The outputs of the most powerful bursts are equivalent to converting the entire mass of the Sun into pure energy in scarcely 10 seconds. For scientists who love explosions, gamma-ray bursts are as exciting as it gets.

Understanding what triggers gamma-ray bursts is one of the foremost challenges in high-energy astrophysics. Theorists have some intriguing ideas. But first some background. We can thank the Cold War for providing us with the satellites that first spotted gamma-ray bursts. The Vela satellites, launched by the United States in the 1960s, were designed to spy gamma rays from clandestine tests of thermonuclear weapons in the atmosphere. But instead of seeing Soviet bombs, the satellites saw one flash after another from space. It took years for military officials to declassify the data. When they did, in 1973, astronomers didn't know how to classify what they saw.

For two decades a debate divided the astrophysics community. Most researchers felt that gamma-ray bursts came from relatively nearby—perhaps from unusual stars in and around our Milky Way. In that case the releases of energy would be impressive but not overly so. A vocal minority maintained that the bursts were fantastically powerful beacons from the depths of the observable universe. Neither camp reached a consensus on how the bursts arose. The confusion was evident in research papers. More than 2,000 were published during those two decades, advancing many notions about how gamma-ray bursts worked.

The logjam broke in the early 1990s with data from a new suite of satellites, including NASA's Compton Gamma Ray Observatory. Measurements of the positions of thousands of gamma-ray bursts showed that they created a random pattern. Their distribution did not follow the hazy band of the Milky Way on the sky, so it seemed unlikely that stars in our galaxy were the culprits. As the evidence mounted, astrophysicists abandoned their biases in favor of the minority view: Gamma-ray bursts indeed were remote in the universe and unimaginably powerful. Near the end of the decade an Italian satellite called BeppoSAX helped solidify that conclusion. BeppoSAX pinpointed the positions of bursts quickly and accurately enough to beam their locations to observatories on the ground. Astronomers who received the messages swung their telescopes to those spots on the sky. For the first time, they saw visible glows from many bursts—occasionally within seconds, with the help of robotic telecopes. The glows, presumably from fireballs accompanying the bursts, often lingered for weeks. The Hubble Space Telescope and large telescopes on the ground followed up by taking images of the glows. Sure enough, they were embedded within faint fuzzy patches: distant galaxies. Spectra of the galaxies revealed that they were hundreds of millions or even billions of light-years away.

Satellite observations also revealed that gamma-ray bursts are as unique as fingerprints. They last anywhere from a few hundredths of a second to a few thousand seconds. Some bursts have multiple peaks of gamma rays, while others have only one.

Enigmatic Cosmic Blasts

The powerful cosmic explosions called gamma-ray bursts are notoriously hard to study. Intense flashes of high-energy radiation, they can happen in any region of the sky without warning. But on January 23, 1999, astronomers from all over the world tracked the visible glow of a gamma-ray burst while it was still emitting high-energy radiation. The blast was the most powerful cosmic explosion ever recorded. For a brief moment the gamma-ray burst rivaled the radiance of 100 million billion stars. Follow-up images zeroed in on the host galaxy (*right*), located two-thirds of the way to the horizon of the observable universe.

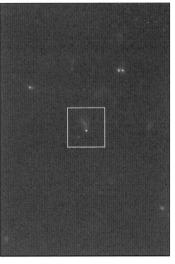

In these Hubble Space Telescope images taken two weeks after the gamma-ray burst, the visible-light fireball has faded to one four-millionth of its original brightness (*left*). An enlargement (*right*) shows the host galaxy as fingerlike filaments extending above the bright white blob of the fireball.

Two neutron stars complete their inexorable merging, beaming powerful jets of gamma rays in opposite directions. Such events may be a source of gamma-ray bursts, enormous outpourings of energy seen throughout the universe. Other gamma-ray bursts may arise from so-called hypernovas, theorized exploding stars displaying exceptional violence.

This variety of energy outputs doesn't make it any easier to figure out what's going on. Imagine seeing fireworks for the first time. You'd be hard pressed to come up with a single explanation for the giant red and green chrysanthemums, the white sparkles that glimmer for many seconds, and the orange squiggles that whistle as they plunge toward you—not to mention those concussive duds that leave your ears ringing. You would probably conclude that more than one explosive mechanism was at work.

The same may be true for gamma-ray bursts. Two popular models have emerged from the pack. To no one's surprise they both involve the densest forms of matter in the universe—the only kind capable of spawning such detonations. In one model two neutron stars whirl around each other in close orbit. We know that such systems can arise if each member of a binary pair of stars blows up in a supernova and leaves behind a neutron-star core. Einstein's general theory of relativity predicts that the two stars will slowly approach each other in a death spiral. The angular momentum lost from the system during this process gets converted into gravitational waves, which speed away from the stars like ripples on a pond. After hundreds of millions of years, the neutron stars will merge. Computer simulations show that the last few seconds of this event may liberate more energy than all the stars in the universe shining during those seconds. Most of that energy will be in gamma rays.

Other researchers think the most powerful gamma-ray bursts of all come not from neutron stars but from black holes, at the instant of their birth. This model calls for an exotic fate to befall certain collapsing stars. If the original star is especially massive and spins especially fast, it may not explode when the core collapses into a black hole. Rather, the model claims, the bulk of the star's gas would spiral into a disk around the black hole at nearly the speed of light. Within about 20 seconds the hole would gulp the entire disk. This ferocious process would heat the gas to 20 billion degrees and shoot stupendous jets of energy out the top and bottom of the disk, where the gas is least dense. Particles within these jets would crash violently into one another and into nearby clouds of gas in space, sparking the bursts of gamma rays. Because "super" lacks enough power as an adjective to describe such chaos, astrophysicists coined the word "hypernova" for a black hole born in this way.

There is some observational evidence that gamma-ray bursts do indeed funnel their energy into tight channels. The jets from gamma-ray bursts are similar to the beams of radiation from a pulsar's poles but narrower and vastly more energetic. If this model holds up, it means that viewers off to the side would not see the gamma-ray burst. Instead, they might see something resembling an "ordinary" supernova. A hypernova may reveal its true nature to about 1 percent of the universe—the portion that lies along the line of sight of either of its two beams.

In the unlikely event that one of these bursts went off in our Milky Way, it probably would sterilize any planets in the path of its beams. Gamma rays are the most energetic form of light. An intense blast of them could wipe out the protective ozone in a planet's atmosphere and expose the planet to deadly radiation from its own star as well as from space. Astronomers learned recently that Earth is not immune to such influences. A lone neutron star in the Milky Way flared up in 1998, spitting out a relatively small burst of gamma rays. By the time the radiation reached Earth it had dwindled to the strength of a dental x-ray. Still, measurements with radio signals showed that the radiation zapped electrons away from atoms in the ionosphere—the uppermost, very tenuous part of the atmosphere. This energetic link between our isolated planet and the rest of the cosmos is fascinating but also vaguely unsettling. When and where will the next powder keg explode?

Where Does the UNIVERSE Go from Here?

To extend motion, matter, and energy to their limits, we must ponder the future of the cosmos. We know that space itself is expanding, carrying galaxies ever farther apart from each other. Will this pattern continue forever? We know that the matter around us is not as solid as it seems and that elements can decay even after billions of years. Can the ultimate constituents of matter survive an infinitely long time? We know that the background energy of the cosmos left over from the Big Bang is slowly winding down, even as colossal bursts still flare up from time to time. Will these displays become ever rarer, until the universe is black and silent?

Existentialists might ponder these questions with delight. They truly put us in our place as inconsequential motes in a vast universe, in terms of both space and time. We would like to know the answers as well for somewhat more practical reasons. Studying the future of the universe requires us to have a firm grasp on its present. To do so we must comprehend many aspects of today's cosmos that elude our limited vision.

One crucial puzzle goes by the name of "dark matter." Simply put, this is stuff we can't see, yet it exerts a gravitational pull like visible matter. Objects that shine may dominate our images of the cosmos, but they hardly make a difference in the big picture of mass in the universe. At least 90 percent of all mass out there is invisible in any wavelength of light—perhaps as much as 99 percent. Indeed, if our telescopes observed gravity rather than light, the cherished galaxies in galaxy clusters would appear as insignificant blips amid giant gravitational fields.

These statements rest on a secure body of observational and theoretical evidence. Inferring the existence of hidden things by looking only at what is visible may sound dubious, but you do it when you see distant headlights coming toward you on a dark road. Even though you cannot see the car itself, you know one is coming. You are familiar with cars, and when you see headlights coming, you know they are attached to a car. In fact, if you are a car buff, you might even deduce how large the car is, what model it is, and how fast it's moving. We also are well aware of the physics of floating ice. A little peak sticking out of the ocean is part of a jagged mass 10 times bigger hidden beneath the waves. Similarly, if we see luminous matter behaving exactly as it would if more mass were present, we confidently predict the existence of that extra mass. Astrophysics buffs can deduce how much mass there is, where it is located, and how it moves.

Among the many signs for dark matter, several convincing ones stand out. Observations of spiral galaxies such as our Milky Way show that their outer stars revolve around the galaxy too quickly for the amount of matter we see. They appear to violate Kepler's laws of orbital motion. The best explanation points to a huge halo of unseen material, contributing mass but no light. The halo extends beyond a galaxy's luminous edge and accelerates the visible matter with its gravitational pull. On larger scales a rich cluster of galaxies must hold tremendous amounts of stars, gas, and dust together with its gravity. But after tallying up all the visible matter in the cluster, we see only a tiny fraction of the mass needed to generate the necessary force. This was discovered in 1936 for the Coma cluster of galaxies by the Swiss-American astronomer Fritz Zwicky. Since then we have confirmed it for every cluster we have observed.

We can thank this mysterious matter for some of nature's most beautiful displays. The collective gravity of an entire cluster can act as a powerful gravitational lens. It warps the light from more distant galaxies into ethereal arcs. Individual galaxies in the cluster also form smaller lenses, so the paths followed by light through these clusters can become quite complex. These cosmic mirages often magnify extremely remote galaxies that we would otherwise miss with our telescopes. In this way, massive gravitational lenses serve as peepholes to the distant universe.

Simulations of the cosmos with powerful supercomputers also point to dark matter's supremacy. Computers now perform so many calculations per second that astrophysicists can explore how the structure of the universe has evolved. Billions of years in the cosmos are compressed to weeks or months of computer time. Many thousands of virtual particles, representing galaxies, drift within a cube that mimics a chunk of the universe measuring hundreds of millions of light-years on a side. The programs simulate the slow but relentless effects of gravity. The attraction draws galaxies together into clumps and weblike filaments, leaving behind gaping voids. But it is the added pull of dark matter that makes these computerized patterns of galaxies resemble the patterns we see in space. Without dark matter, galaxies in the simulations simply fly apart from one another because of their own random motions.

If we have convinced you by now that dark matter exists, you probably are wondering what it is. Join the club. It's the astrophysical mystery of the century—both the twentieth century and, perhaps, the twenty-first. One researcher compared studying dark matter to assembling the pieces of an all-black puzzle in a black room. With thick black gloves on, we might add. Not only can't we see dark matter, but there doesn't seem much hope of putting our hands on it.

This is not to say that progress is impossible. The search for dark matter is happening on many fronts. For a time it seemed that much of the missing matter might be ordinary stuff like stars and planets, but too far away or too dim for us to see. We know that stars below a certain size glow faintly or fail to ignite nuclear fusion in their cores at all. We call such objects brown dwarfs. Jupiter-sized planets could wander the galaxy as well, unaffiliated with any star. Astronomers who envisioned a cloud of these things around the Milky Way dubbed them MACHOs, an aggressive acronym for "massive compact halo objects." How might we detect these ghosts? It turns out that a

Visualizing the Invisible

Stars and galaxies make up less than 10 percent of the matter in our universe; mysterious "dark matter" makes up the rest. It cocoons every galaxy, including our Milky Way. Its immense gravitational pull also controls the growth and evolution of giant groups of galaxies.

Although astronomers cannot detect the dark matter with telescopes, they can infer its distribution through gravitational lens mass tomography, a technique that relies upon the bending of light by strong concentrations of mass.

The gravitational lensing of an extremely distant blue galaxy by the foreground galaxy cluster CL0024+1654, itself 2 billion light years from Earth, results in multiple images (*right*). Powerful gravitational fields within the cluster force the distant galaxy's light to travel along many different paths on its way to Earth, creating the smears of blue light in this Hubble Space Telescope image.

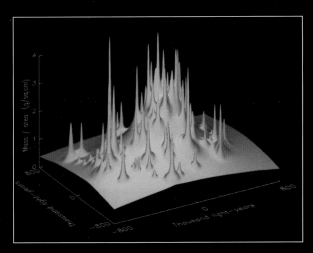

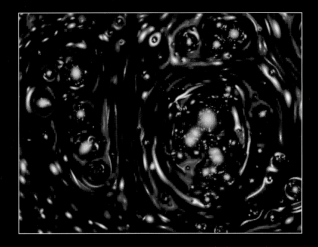

Computer simulations allowed astronomers to retrace the paths that rays of light follow through CL0024+1654, just as tomographic scans reveal the patterns of bone and tissue within a human body. The simulations reveal that only a particular distribution of dark matter within the cluster would create the exact blue patches of light in the Hubble image. The orange peaks represent the most intense concentrations of dark matter in the cluster. Astronomers tested more than one million possible distributions of dark matter to derive this solution.

Astronomers run simulations to test whether dark matter actually resides in tight clumps around individual galaxies. A lumpy distribution of dark matter would create wildly distorted images, in which the blue light from the distant galaxy bends into surreal swirls and loops rather than the smears we see in the Hubble image. Because no such patterns are seen in space, astronomers conclude that most dark matter spreads out fairly evenly within giant clusters and superclusters of galaxies.

The Evolution of Structure in the Universe

Studies of the chilly microwave glow that fills all of space reveal that the early universe was a nearly smooth blend of matter and energy. For many millions of years after the first atoms formed, this almost featureless young cosmos lacked stars and galaxies. But the visible universe today contains more than 100 billion galaxies, arrayed through space in gigantic clusters that seem to form spidery webs of matter. How did they arise?

Theories suggest that tiny gravitational differences from place to place in the infant universe started a snowballing process, gradually pulling matter together into individual stars, galaxies, and clusters of galaxies. For objects the size of galaxies and larger, the powerful gravitational pull of dark matter played the key role in forging the structures that we see in the cosmos today.

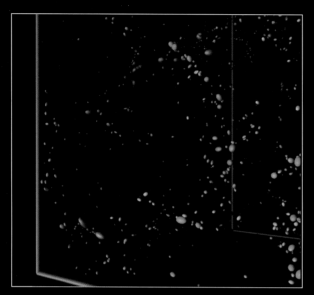 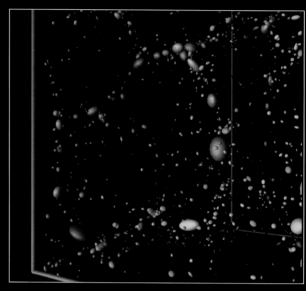

Cosmologists create three-dimensional simulations on supercomputers to mimic the development of structure in the universe, a process that took billions of years in real time. These two sections of the simulation show the same region of the universe across a span of 8 billion years. The colorful blobs represent groups of galaxies and their associated halos of dark matter. Red blobs contain the most massive concentrations of galaxies; blue ones are the least massive shown. When the universe was about 5 billion years old (left), large structures had begun to form under the relentless long-range influence of gravity. Today, 13 billion years after the Big Bang (right), the simulation shows chains and sheets of galaxy clusters separated by enormous voids.

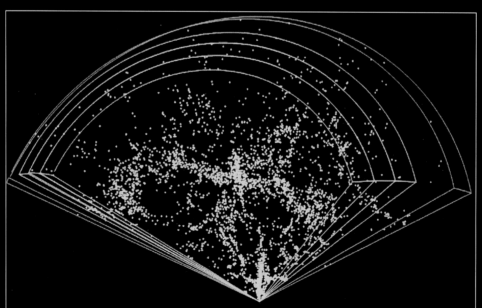

This telescopic survey of the locations of galaxies in the universe today reveals that they do indeed form clusters and sheets in space. The yellow dots are thousands of galaxies extending into space for a few hundred million light-years, with Earth at the bottom point of the wedge. Our once-smooth universe is now riddled by vast clumps and voids.

MACHO passing directly in front of a distant star creates a tiny gravitational lens. We see the star's light spike upward brightly for a few hours or days. These events are so rare that special telescopic systems must monitor millions of stars at once to catch one in action. As technology improves, astronomers are seeing a growing number of minilensing events. However, the jury is still out on what's causing them.

Even if MACHOs are common, they might add just a percent or two to the total mass of the universe. Far more important, most astrophysicists believe, are the contributions of tiny particles. Neutrinos are a strong choice. These flecks, members in good standing of the standard model, stream from all nuclear reactions. That includes fusion in the core of the Sun and the collapse of matter in supernovas, which unleash neutrinos in whopping numbers. Long presumed to be massless, neutrinos may possess the barest whisper of mass, according to recent experiments. The evidence comes from detecting subtle changes in a few neutrinos as they pass through Earth. A tiny fraction oscillates between two types of matter described by the standard model. That transformation is possible only if a neutrino carries some heft—but, mind you, perhaps 100,000 times less than electrons. Even with such tiny scale readings, all the neutrinos in the universe might outweigh the total mass of stars and galaxies.

Other dark-matter candidates have earned the dubious honor of being called WIMPs, for "weakly interacting massive particles." String theory predicts supersymmetric particles as partners to every bit of matter and every force-carrying particle in the universe. If that's true, those "sparticles" (as they are known) exist all around us. They may begin to interact with the matter we know only at very high energies, perhaps beyond the reach of existing accelerators. It's a compelling goose chase. Many physicists suspect that much of the dark matter will consist of completely unknown particles, thus extending the standard model in new and exciting ways.

All this fuss about dark matter is driven by two factors. One simply is our insatiable curiosity about the nature of our universe. Ignorance of 99 percent of the cosmos doesn't sit well with most people, especially astrophysicists. The other is our obsession with fate. And dark matter, like it or not, controls our fate. Specifically, the amount of matter in the universe determines whether space will continue to expand forever or collapse back in on itself from the force of gravity. We call these

Instead, they discovered that the universe seems to be slowing less quickly than gravity should allow—and may even be speeding up with time. It's as if some strange force is counteracting gravity.

eventualities the "open" and "closed" universes, respectively. An open universe continues onward much as our universe today for many billions of years, until everything simply burns out. The cosmic fuel tank of hydrogen, after all, is exceedingly large but not infinite. A closed universe ends in an all-encompassing smash-up of matter—an inverse Big Bang that we could name the "big squeeze." Such an event might even lead to a new Big Bang, but nothing you recognize today would survive the transition from one universe to the next.

We'll eliminate the suspense: It looks more and more likely that we live in an open universe. For years the debate raged about whether there is enough mass to slow and eventually halt the Hubble expansion. The latest tallies have shown that the answer is *no*. All the visible matter and dark matter in the universe appear to add up to no more than 30 to 40 percent of the amount needed. Whatever process forged our supply of matter in the Big Bang clearly did not do so with the intent of making the universe fall back into a fiery point.

Recently, another curious finding made it appear even more likely that our cosmos faces an expansive future. Certain supernova explosions serve as excellent "standard candles" in the universe. Like a succession of 100-watt lightbulbs on a row of porches, these supernovas nearly match one another in luminosity no matter where they pop off. It's then straightforward to figure out how far away they are. If one supernova is nine times fainter than another, it's about three times farther away, and so on—the simple "distance squared" formula for brightness that also applies to shining objects on Earth. In this way researchers found that they could gauge the distances to many galaxies scattered throughout the universe by watching for supernova blasts. (One goes off somewhere in the universe every second, so large telescopes detect supernovas regularly.) Then, they could measure the speed at which each supernova's host galaxy is moving away from us by studying the galaxy's spectral lines.

When combined, these observations reveal a chronology of the universe's expansion. Because of the effects of gravity, the astronomers had expected to see that the expansion rate is gradually slowing as time passes, but not enough to ever stop completely. Instead, they discovered that expansion of the universe seems to be slowing less quickly than gravity should allow—and may even be speeding up with time. It's as if some strange force is counteracting gravity. What's going on?

This may ring a bell for Einstein aficionados. When Einstein wrote his general theory of relativity in 1916, scientists presumed the universe was static. From his equations Einstein derived the existence of a repulsive energy that acts like an anti-gravity force. Such a force, he said, would balance out gravity to keep the universe at a stable size. But when Hubble discovered the outward motions of galaxies a decade later, that need vanished. Einstein quickly discarded the antigravity addition to his equations and called it his greatest mistake. Now it appears that he may have been right all along. Many astrophysicists favor the repulsive energy, called the cosmological constant, as the best explanation of the supernova data. It is indeed tempting to think of it as a literal "antigravity." But in fact the cosmological constant has nothing whatsoever to do with matter. It is a springiness inherent in space itself, an outward pressure that grows as space expands. The more space there is, the springier it becomes. In other words, if the cosmological constant is real, it will force the universe to expand faster and faster without limit. Quantum mechanics may explain the force as a "vacuum energy" present throughout the void of space, but a consensus does not yet exist.

All of this research allows us to paint a portrait of our future in the cosmos. Let's optimistically assume that our civilization endures for a billion years. If our descendants haven't found a way to colonize other planetary systems by then, they'll be out of luck. The Sun will gradually start to brighten, sterilizing Earth with increasing radiation. In about 5 billion years the Sun will use up its hydrogen fuel and swell into a red giant. Earth may escape being swallowed by the Sun's outer atmosphere, a 3,000-degree plasma, because the Sun will shed a great deal of mass. This will force Earth's orbit to slowly move outward in the solar system. Even so, the oceans will boil off, the atmosphere will evaporate, and the crust itself may melt. Earth will be a charred ember.

Another event shortly thereafter could stir up havoc elsewhere in the Milky Way. The Milky Way and our sister galaxy in Andromeda are moving toward each other at the leisurely clip of 250,000 miles per hour. We may collide in 6 billion years or so. We don't yet know enough about the sideways motion of the Andromeda galaxy relative to that of the Milky Way, but it could be a direct hit. Individual stars are so far apart that they aren't likely to collide in such an encounter. However, close flybys between stars would disturb their giant clouds of comets, scattering many of them like leaves in the wake of a passing truck. Comets could bombard any planets orbiting the stars for millions of years. The gravitational mayhem could fling some planets from their solar systems entirely. Living in the Milky Way during that era will require adaptability and a ready fleet of interstellar ships, to be sure.

To gaze into an even bleaker future, we must stop being parochial and let our minds wander beyond the Milky Way. We also need to think in units of time that don't enter most dinner conversations. The smallest stars burn their nuclear fuel so slowly that they will survive for about 10 trillion years. By that time galaxies will start shutting down their star formation as supplies of hydrogen dwindle. Galaxies also will be extraordinarily far apart by then as the expansion of the universe accelerates. After 100 trillion years, no more stars will shine. Black holes, neutron stars, and white dwarfs, long since cooled to blackness, will drift within the dark remains of galaxies.

How Protons Decay

Protons seem like bastions of stability in the nuclei of atoms. However, theories predict that all protons will disintegrate into tiny waste products far in the future, perhaps a trillion trillion trillion years from now. Physicists are watching for these ultrarare decays, but so far they have eluded detection.

According to one theory, a decaying proton unleashes a predictable blizzard of exotic and short-lived particles. The strong nuclear force that binds the three quarks composing a proton (1) breaks down, allowing two quarks (2) to spontaneously merge into a massive particle called a leptoquark (3). This unstable particle ejects two flecks of antimatter: a positron (4) and an antiquark (5). The antiquark merges with the remaining quark (6) from the former proton to create yet another transient particle, a pi zero (7). Finally, this particle disappears in a flash of gamma-ray photons (8). These high-energy rays of light could form pairs of electrons and positrons (9), which might destroy each other and release still more photons. These fleeting bits may be all that remains when solid matter as we know it vanishes from the cosmos.

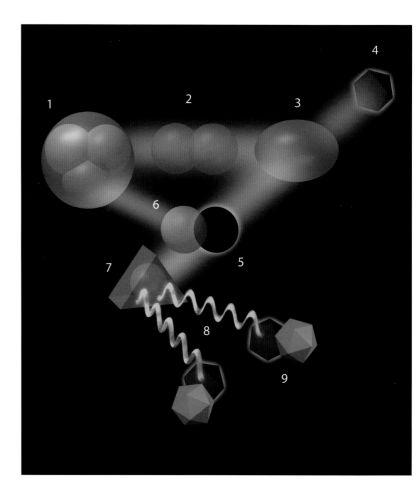

Occasional collisions will spark spectacular flashes of light. Then, after about a trillion trillion trillion years, matter itself will decay. Current theories predict that protons are unstable on these timescales. When they vanish into blips of gamma rays and tiny waste products of matter, nothing made of atoms will remain.

The sole denizens of this dark universe will be black holes. But they too disappear, in the process called Hawking radiation—a quantum-mechanical annihilation of particles that slowly evaporates a black hole and emits gamma rays. The largest black holes of all, those at the centers of galaxies today, will take a staggeringly long time to evaporate: a googol, or 1 followed by 100 zeros, years. When the last black hole shrinks and destroys itself in a bright burst of gamma rays, we may then declare that the universe has died—not with a whimper, but a bang.

These fantastic outcomes are all grounded in scientific knowledge that generations of astronomers and physicists have assembled with care. Yet when we look to the future or gaze deep into space, we still face profound limits to our knowledge. Those barriers are erected by our narrow abilities to imagine how the universe works. Two millennia ago astronomers believed that there were just 2,000 stars in all the heavens. They were all equally far away from Earth, it seemed, placed on a large spherical shell that marked the boundary of the cosmos. Could Ptolemy, laboring to understand the universe in the second century A.D., ever have imagined that there are 100 billion stars in our galaxy alone, all at different distances? Could he have grasped 100 billion other galaxies beyond the Milky Way?

Today, we labor to answer the great astronomical questions of our own era. Thousands of years from now astronomers may smile back at us as they learn about our futile efforts to fit the universe into a model barely resembling a larger reality. We hope we are not completely off track. But even if we are, we build the lower rungs of the ladder of astronomical knowledge for future generations—just as Ptolemy did for us 18 centuries ago. Ultimately, the greatest progress will come when we realize that the questions we must answer are the ones we have yet to ask.

What Lies Ahead

The amount of dark matter in space is a key factor that controls the fate of our expanding universe. Cosmologists describe three possible fates, illustrated here as simple shapes that depict the curvature of the space-time fabric of the cosmos. If enough dark matter existed to raise the average density of matter in the universe to three atoms of hydrogen per cubic meter of space, space-time would be flat (*left*). The relentless pull of this matter's gravity would exactly oppose the force of expansion, creating a universe that gradually slows down but never quite stops expanding. More matter than this would result in a victory for gravity, leading to a closed universe (*center*) that collapses back in on itself after many billions of years. However, astronomical evidence shows that the universe contains just 30 to 40 percent of the matter needed to counteract its expansion. As a result, most cosmologists believe we live in an open universe (*right*) that will expand forever. A curious repulsive force within the fabric of space itself may even cause the cosmos to grow ever more rapidly as time passes.

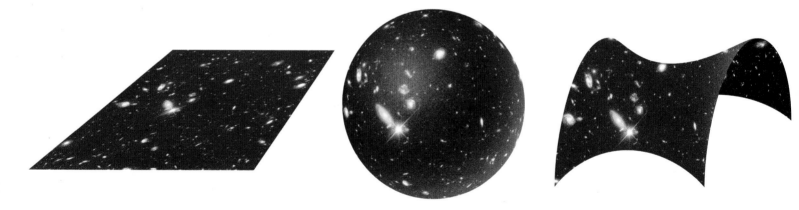

The Ultimate Heat Wave

The Sun is a mellow middle-aged star about halfway through its expected 10-billion-year lifetime. As the star continues to age, however, it will consume hydrogen more quickly in order to maintain its internal temperature and pressure. The extra work will cause the Sun to expand and grow bigger, brighter, and hotter. In about 4 billion years, Earth will experience a global warming beyond even the direst of today's predictions concerning greenhouse gases. When the Sun turns to helium as a fuel source in its core, it will swell even further into a red giant looming over the doomed Earth's horizon.

A Dim Future for Galaxies

Galaxies will produce new generations of stars as long as their stores of hydrogen last. However, this fuel tank is not infinite. As stars continue to process hydrogen into helium and heavier elements, fewer stars will form. The hydrogen supply will dwindle after about 10 trillion to 100 trillion years, and galaxies in the cosmos will gradually fade. The longest-lived stars in the universe, tiny red dwarfs that ration their nuclear fuel, also will shine dimly about that long. Thereafter, the sole denizens of galaxies will be ultracompact objects: cooling white dwarfs, neutron stars, and black holes. Collisions between these bodies may release bursts of light, but by and large, galaxies will be invisible.

Fade to Black

Astronomers who have gazed into the very distant future see a dark universe indeed. About one googol—1 followed by 100 zeros—years from now, the last black holes will evaporate through Hawking radiation. After each black hole shrinks to the size of an atom, it will vanish in a flash of gamma rays that briefly illuminates the darkness (*upper right*). Then the cosmos will consist of isolated photons, electrons, positrons, and neutrinos (*foreground*), drifting through space and separated by immense distances comparable to the size of our visible universe today.

Progress in Understanding the Cosmos: A Selected Chronology

ca. 440 B.C.

Greek philosopher Leucippus and his student Democritus propose that dividing a piece of iron in half again and again eventually leads to "atoms," from the Greek word for "indivisible." These basic units compose all matter, the philosophers reason.

ca. A.D. 140

Claudius Ptolemy, a Greek astronomer in Alexandria, Egypt, constructs a complex Earth-centered model of the universe. His model places the planets on small circles, called epicycles, that revolve around Earth on larger spheres. The resulting curlicues in planetary motion appear to explain the occasionally erratic wanderings of the planets in the sky.

1572

Danish astronomer Tycho Brahe spies a bright new star: a supernova in the constellation Cassiopeia, showing that the realm of the stars is not permanent and unchanging. Brahe then establishes an observatory and assembles an exhaustive record of the motions of planets and comets in the sky. The work prepares the way for the discoveries of his assistant, Johannes Kepler.

1610

Italian physicist and astronomer Galileo Galilei publishes the first observations of the night sky through a telescope in *The Starry Messenger*. His discoveries of moons orbiting Jupiter and the phases of Venus support the Copernican view of a Sun-centered solar system, but the Catholic Church forces Galileo to recant his claims. Galileo also lays the foundation for modern physics with his studies of moving objects on Earth.

1687

English physicist Sir Isaac Newton publishes *The Mathematical Principles of Natural Philosophy*. It contains his three laws of motion, still taught in physics classes today, and his law of universal gravitation, which explains the motions of falling objects and planets in orbit as the results of a single force. Newton is also known for his investigations of the nature of light, using prisms.

ca. 260 B.C.

Aristarchus of Samos, Greece, becomes the first philosopher to suggest that Earth orbits around the Sun within a giant universe. However, his writings on this topic do not survive; we have only allusions to his work.

1543

Polish astronomer Nicolaus Copernicus publishes his life's work in *On the Revolutions of the Celestial Spheres.* His treatise describes his conviction that the Sun, not Earth, rests at the center of the universe. Copernicus receives a copy of the printed book on his deathbed and his contributions are not recognized until many years later.

1609

German astronomer Johannes Kepler publishes his first two laws of planetary motion in *The New Astronomy*. The first and most famous law states that the orbits of planets trace ellipses in space, not circles, as astronomers had assumed. The second law describes how planetary orbits circumscribe equal areas in equal amounts of time. His third law, which describes how the length of a planet's year depends on the size of its orbit, appears 10 years later in *The Harmonies of the World*.

1655

Dutch astronomer Christian Huygens realizes that the "ears" of Saturn, as Galileo had called them, actually form a ring that encircles the planet without touching its surface.

1820

Danish physicist Hans Christian Oersted shows that an electric current moving through a wire generates a magnetic field around it. This first convincing demonstration of a link between electricity and magnetism sets the stage for much of nineteenth-century physics.

1831

English physicist Michael Faraday, regarded as the father of the electrical age, proves the reverse of Oersted's discovery: A magnet moving through a coiled wire generates an electric current in the wire. Faraday uses this discovery to build the first dynamo, or electric generator.

1843

English physicist James Joule establishes the equivalence of work and heat in a series of experiments covering more than three decades. His inventions include a famous apparatus in which the gravitational potential energy of a falling weight converts to thermal energy within a jar of water via a set of rotating paddles. The principle of the conservation of energy stems largely from Joule's work.

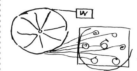

1873

Scottish physicist James Clerk Maxwell publishes his unified theory of electromagnetism, in which light waves arise from electric fields and magnetic fields that stream through space. The fields oscillate together like water waves and travel at a fixed speed. Maxwell's work forms the basis for Albert Einstein's special theory of relativity.

1901

Italian physicist Guglielmo Marconi sends the first wireless transmission across the Atlantic Ocean. His simple Morse Code signal presages the age of radio communication that dominates our technological society.

1905

German physicist Albert Einstein derives one of his landmark advances in physics, the special theory of relativity. It asserts that light

always moves at a fixed speed regardless of the motions of an observer, a tenet with strange consequences for measurements of mass, length, and time. His monumental general theory of relativity, published in 1916, describes how gravity, matter, and space interact throughout the cosmos.

1842

Austrian physicist Christian Doppler observes that waves of sound or light from a moving object are compressed if the object moves toward an observer and stretched if the object moves away. This effect shifts either the frequency of the sound waves or the wavelength of the light in a precise way. Astronomers use the Doppler shift today to track the motions of stars and galaxies in the universe.

1858

American astronomer Henry Draper pioneers astronomical photography by mounting photographic plates on the back of a reflecting telescope. In 1872, Draper takes the first successful photograph of the spectrum of a star, in which a prism spreads the star's light into its component colors.

1900

German physicist Max Planck proposes that energy radiates not in a smooth and continuous spectrum but in tiny packets called quanta. This insight sets into motion a true revolution in physics, called quantum mechanics. Many physicists develop this nonintuitive theory of the behavior of matter and light during the next three decades.

1887

American physicists Albert Michelson and Edward Morley find that light moves at a constant speed, unaffected by Earth's motion through space. This finding, which disproves the notion of an invisible "ether" through which electromagnetic waves travel, also plays a key role in Einstein's theories.

1909

Physicist Sir Ernest Rutherford, a New Zealand native working in England, determines that most of an atom's mass resides in a tiny positively charged nugget at its center, surrounded by a flitting cloud of electrons. The startling implication of Rutherford's work is that matter consists mostly of empty space.

1912

American astronomer Henrietta Swan Leavitt unveils a relationship between the brightness and the flickering rate of certain pulsating stars called Cepheid variables. The advance allows future astronomers, most notably Edwin Hubble, to gauge the distances to faraway galaxies by measuring these cosmic lightbulbs.

1916

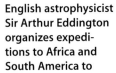

German astrophysicist Karl Schwarzschild uses Einstein's general theory of relativity to calculate how small an object would have to shrink for its gravitational field to become so intense that light could not escape. This Schwarzschild radius is about 5 miles for a star with three times the mass of our Sun. Schwarzschild's work forms the theoretical basis for our understanding of black holes.

1919

English astrophysicist Sir Arthur Eddington organizes expeditions to Africa and South America to look for small shifts in the positions of stars near the Sun during a total eclipse. He sees displacements exactly as predicted by the general theory of relativity. Eddington also is the first to propose that hydrogen atoms fusing into helium in the Sun's core might provide the source of its energy.

1929

American astronomer Edwin Hubble discovers that the universe is expanding in all directions. His analysis is based on the distances to 25 galaxies which he and American astronomer Vesto Slipher determined from Cepheid variables and on their rates of motion away from Earth, determined from their Doppler shifts. This discovery remains one of the best pieces of evidence in support of the Big Bang theory.

1936

Swiss-American astronomer Fritz Zwicky observes that galaxies in the Coma cluster appear to move much more quickly than could be explained by the gravitational attraction of the galaxies alone. The cluster contains about 10 times as much invisible matter as visible matter, Zwicky concludes. This "missing mass" problem lurks in the background for 40 years but has now emerged as one of the central mysteries of astronomy.

1913

Danish physicist Niels Bohr proposes a new model for the hydrogen atom. The atom's electron can occupy only a set of specific orbits, or energy levels, around the nucleus, Bohr claims; all other orbits are forbidden. Although the details of this model change over the years, it explains the distinctive patterns of light emitted by excited atoms.

1925

Austrian physicist Wolfgang Pauli derives a quantum-mechanical rule now known as the Pauli exclusion principle. It states that no two electrons can inhabit the same location with the same motions at the same time. Pauli's principle explains the previously mysterious order of the periodic table of the elements.

1927

German physicist Werner Heisenberg publishes his famous uncertainty principle. One cannot simultaneously know the exact position and the exact momentum (velocity times mass) of a particle at any given moment, the principle maintains. This strange notion became a cornerstone of quantum mechanics.

1950

American astronomer Fred Whipple creates his dirty snowball model for the nucleus of a comet. He envisions a chunk of ice mixed with grains and dust left over from the birth of our solar system. Whipple's model accounts well for the overall appearance of comets, including their ethereal tails of gas and dust released under the heat of the Sun.

1948

American physicist Ralph Alpher and Russian-born physicist George Gamow publish the first model

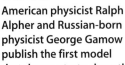

that demonstrates how the elements arose in the early universe. They envision a hot and dense fireball that forges hydrogen, helium, and a smattering of other elements. Gamow also calculates that this inferno of creation left behind a faint glow that suffuses the universe today, a prediction verified by Arno Penzias and Robert Wilson in 1964.

$$N = R_* \, f_p \, n_e \, f_l \, f_i \, f_c \, L$$

1961

American astronomer Frank Drake devises a formula to estimate how many civilizations elsewhere in our Milky Way galaxy are capable of communicating over long distances in space. The formula, now known as the Drake equation, requires astronomers to calculate a string of challenging probabilities, such as the odds that intelligent life arises on a planet. Thus far, we know of only one such civilization: our own.

1967

English radio astronomers Jocelyn Bell and Antony Hewish discover steady signals from the first known pulsar, a rapidly rotating neutron star that sweeps the galaxy with beams of radio waves and other forms of light. Astronomers have since found

more than 1,000 of these dense stellar cinders, which can spin hundreds of times every second.

1981

American astrophysicist Alan Guth proposes that the infant universe underwent a period of extraordinarily rapid expansion, called inflation, for less than one billion trillion trillionth of a second. This expansion explains the smoothness of the cosmos on the largest scales. Inflation also suggests that galaxies and clusters of galaxies condensed from quantum fluctuations in the fabric of space-time during the Big Bang.

1987

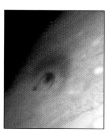

Canadian astronomer Ian Shelton, working at an observatory in Chile, finds an unexpected flare of light on a photograph of the Large Magellanic Cloud, a nearby galaxy. The new "star" is Supernova 1987A, the brightest supernova seen from Earth in 383 years. Astronomers study the explosion to learn new details about the deaths of giant stars.

1993

An American team of planetary scientists, Eugene and Carolyn Shoemaker, and astronomer David Levy find a comet captured by Jupiter's gravity. Astronomers watch in 1994 as Comet Shoemaker-Levy 9, broken up into a long chain of fragments, plows into Jupiter in the first recorded collision of objects in our solar system. The explosive impacts create dark Earth-sized plumes of gas in Jupiter's atmosphere.

1965

American physicists Arno Penzias and Robert Wilson find a faint glow that pervades the universe: the cosmic microwave background radiation. This glow, which fills space at a temperature of less than 5 degrees Fahrenheit above absolute zero, is strong evidence that the entire universe has cooled down uniformly since its birth in the Big Bang about 13 billion years ago.

1964

American physicists Murray Gell-Mann and George Zweig propose that the protons and neutrons in an atomic nucleus are each made of a trio of still smaller particles, fancifully named quarks by Gell-Mann. Today, physicists view quarks, electrons, and neutrinos as the basic constituents of matter.

1974

English physicist Stephen Hawking determines that black holes emit particles and slowly evaporate over time, a quantum-mechanical process now known as Hawking radiation. As a result, in the distant future of the universe all black holes will shrink and ultimately explode when they become very small.

1995

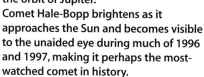

Americans Alan Hale, a professional astronomer, and Thomas Bopp, an amateur astronomer, independently discover a large new comet beyond the orbit of Jupiter. Comet Hale-Bopp brightens as it approaches the Sun and becomes visible to the unaided eye during much of 1996 and 1997, making it perhaps the most-watched comet in history.

1990

American planetary scientist Carl Sagan and his colleagues use the *Galileo* spacecraft to test whether a space probe can find unambiguous signs of life on a planet—in this case, Earth—without landing on its surface. Sagan also led a team of astronomers and artists who created gold-plated phonograph records containing sounds and images from Earth, to travel aboard the *Voyager* spacecraft far beyond our solar system.

Glossary

Atom
Smallest unit of matter for which a chemical element retains its identity. Composed of protons and neutrons in a compact nucleus surrounded by a cloud of electrons.

Big Bang
Theory that an explosion 13 billion years ago created all matter and energy in the universe. Observed expansion of the cosmos and a faint glow of microwave light that fills space support the theory.

Black hole
A region of space with a gravitational field so intense that the fabric of space curves back upon itself, preventing everything, including light, from escaping. Born during the collapse of very high mass stars and often within the matter-rich centers of galaxies.

Blue shift
Shortening of the wavelength of light as a radiating object and an observer move toward each other.

Cepheid variable
Pulsating star that flickers regularly at a rate that depends on its luminosity. Used to gauge distances to other galaxies; played a key role in the discovery that the universe expands.

Conservation of energy
Principle that the total amount of energy within a system remains constant unless an outside force acts upon it. Disguised when energy shifts from one form to another, including kinetic (the energy of motion), potential (the capacity of an object to move, such as under the influence of gravity), and heat.

Conservation of momentum
Principle that the total amount of momentum (an object's mass times its velocity) remains constant for a system of objects unless an outside force acts upon it. Related is the concept of conservation of angular momentum, which takes into account the spins and orbital motions of objects.

Copernican principle
Proposition credited to Nicolaus Copernicus that Earth and its inhabitants exist in an ordinary place and time rather than at the center of the cosmos.

Coriolis effect
An apparent force acting on objects moving north or south across the surface of a spinning body, caused by east-west motion that is faster near the equator than near the poles.

Cosmic microwave background
Faint radiation, principally in microwaves and radio waves, that pervades the universe at just a few degrees above absolute zero. Represents leftover warmth from the Big Bang.

Cosmological constant
A part of Einstein's equations that calls for a repulsive pressure that may arise from the springiness of the vacuum of space, counteracting gravity and causing the universe to expand more quickly with time.

Cosmological principle
Postulate that, on average, large-scale properties of the universe are the same everywhere and that the same laws of physics apply throughout the cosmos.

Dark matter
Invisible and unknown material that accounts for more than 90 percent of the gravity in the universe. Has pronounced effects on the formation and evolution of galaxies and clusters of galaxies.

Doppler effect
Change in frequency of waves as an object and an observer approach or recede from one another. Used to gauge how quickly celestial bodies move.

Electromagnetism
One of four fundamental forces of nature, arising from simultaneous motions of electric and magnetic fields through space. Responsible for attraction and repulsion between charged and magnetized objects, and for propagation of light waves.

Entropy
Relentless tendency of an isolated system of objects to become increasingly disordered with time.

Event horizon
Boundary around a black hole marking the zone from which light can no longer escape the hole's gravitational pull.

Gamma-ray burst
Titanic release of energy, especially high-energy gamma rays, from massive objects in distant parts of the universe. Their origin remains a mystery, but they are thought to arise from the birth of black holes in large supernovas or from collisions of neutron stars.

Gravitational lensing
Bending of light along curved or multiple paths through space as a result of the gravity of a massive object, such as a star or a cluster of galaxies, lying between an observer and a distant light source.

Gravitational wave
Subtle ripple in the four-dimensional fabric of space-time caused by the sudden motion of a massive object. Moves through space at the speed of light.

Gravity
Attraction between two objects based solely on their mass and the distance between them. Although the weakest by far of the four fundamental forces of nature, it extends over the greatest distances.

Greenhouse effect
Warming of the surface and lower atmosphere of a planet by a blanket of carbon dioxide, water vapor, and other gases that prevent infrared energy from escaping into space.

Habitable zone
Region around a star in which liquid water can exist on or beneath the surface of a planet or moon, thereby providing a possible habitat for life as we know it.

Hawking radiation
Quantum-mechanical process which enables black holes to evaporate slowly over time until they vanish in a burst of gamma rays and subatomic particles.

Inertia
Tendency of a moving object to keep moving in a straight line until some external force, such as friction, makes it change. Equivalently, the tendency of a stationary object to remain at rest until an external force acts upon it.

Inflation
Hypothesis that the universe expanded

exponentially for a tiny fraction of a second immediately after the Big Bang. May have smoothed the cosmos to the uniformity we see today while imprinting subtle fluctuations that led to galaxies and clusters of galaxies.

Magnetosphere
Region of space surrounding a planet in which the planet's magnetic field deflects the solar wind, shielding the planet from charged particles.

Neutrino
Fundamental particle produced by radioactive decay, fusion in the cores of stars, and energetic events such as supernova explosions. Contains little or no mass and rarely interacts with other matter.

Neutron star
Ultracompact object left behind at the core of many supernovas. Consists entirely of neutrons and packs slightly more than the mass of the Sun into a sphere about a dozen miles across.

Nucleosynthesis
Creation of elements within the Big Bang (mostly hydrogen and helium), the interiors of stars (elements as heavy as iron), and supernovas (all heavier elements).

Periodic table
Chart that arranges the known chemical elements in rows and columns according to their properties, which arise from the fundamental rules of quantum mechanics.

Photon
Energy-carrying particle of electro-magnetism with no mass, acting simultaneously as a particle and a wave. Commonly thought of as the basic particle of light.

Plasma
Charged gas in which electrons or ions (atoms that have lost or gained electrons) can move freely, carrying electric currents through space.

Precession
Change in the direction of the spin axis of a rotating body caused by force that acts on an equatorial bulge or other nonspherical aspect of the body. Makes Earth's axis trace a cone in space once every 26,000 years as a result of combined gravitational tugs from the Sun and the Moon.

Pulsar
Rapidly spinning neutron star that emits beams of radiation on sweeping paths through space. Usually detected in radio waves.

Quantum mechanics
Twentieth-century theory of matter and energy that describes light and elementary particles as both particles and waves. Introduces uncertainty and probability into physics by limiting electrons to certain "energy levels" around atoms and preventing physicists from simultaneously measuring the position and momentum of any particle.

Quark
Basic constituent of matter that makes up protons and neutrons in sets of three and other subatomic particles in sets of two.

Quasar
Nuclei of some galaxies near the fringes of the observable universe that emit powerful streams of x-rays, radio waves, and visible light. Probably powered by supermassive black holes consuming nearby stars and gas.

Radioactivity
Spontaneous decay of an unstable atom into another atom through the capture or emission of subatomic particles.

Red giant
Bloated end-stage of the life of a star as it consumes helium within its core and sheds its outer layers into space.

Red shift
Lengthening of the wavelength of light as a radiating object and an observer recede from each other.

Relativity, general
Extension of Albert Einstein's special theory of relativity to include the effects of acceleration and gravity. Explains gravitational attraction as dimples in the four-dimensional fabric of space-time around massive objects.

Relativity, special
Theory advanced by Einstein in which a beam of light is measured to move at a constant speed regardless of the motion of the observer. Results in altered measurements of time, length, and mass for a rapidly moving object relative to a stationary observer.

Solar wind
Stream of high-energy charged particles that blow outward through the solar system from the Sun's atmosphere.

Standard model
Current model of the universe that serves to account for the behavior and interaction of all known particles. Holds that the basic units of matter are quarks, electrons, and neutrinos, which interact through gravity, electromagnetism, and the strong and weak nuclear forces.

String theory
Candidate for a "theory of everything" that combines all nature's basic forces into a coherent description of the universe. Depicts fundamental particles as tiny vibrating loops, not points.

Strong nuclear force
Fundamental force of nature that binds protons and neutrons within atomic nuclei. Strongest of all forces, but acts only at subatomic distances.

Sublimation
Direct transformation of a solid into a gas, as in the tail of a comet.

Supernova
Cataclysmic explosion of a massive star that has started to accumulate iron in its core, triggered by a shock wave when the outer layers of the star rush inward. Forges elements heavier than iron and blasts them into space.

Tides
Distortions raised in the body of a celestial object by an outside gravitational field which pulls with different strengths on the near and far sides of the first body.

Weak nuclear force
Fundamental force of nature that mediates the radioactive decay of atomic nuclei.

White dwarf
Compact Earth-sized remnant left behind by the collapse of a star like the Sun after it consumes the nuclear fuel in its core.

About the Authors

Neil de Grasse Tyson, an astrophysicist, is the Frederick P. Rose Director of the Hayden Planetarium at the American Museum of Natural History in New York City. He is also Visiting Research Scientist at Princeton University. Since 1995 Tyson has written monthly essays for *Natural History* magazine under the title "Universe." Tyson's most recent books include *Just Visiting This Planet*, a playful question and answer book about the universe, and his memoir *The Sky Is Not the Limit: Adventures of an Urban Astrophysicist*. He was born and raised in New York City and is a graduate of the Bronx High School of Science. Tyson went on to earn a B.A. in physics from Harvard and a Ph.D. in astrophysics from Columbia.

Charles Tsun-Chu Liu is an astrophysicist at the American Museum of Natural History in New York City and Visiting Research Scientist at Columbia University. His research focuses on galaxy collisions, starburst galaxies, and the star formation history of the universe. Liu's work has appeared in numerous scientific publications, including *The Astrophysical Journal* and *The Astronomical Journal*. Liu earned his B.A. in astronomy and astrophysics and in physics at Harvard and his Ph.D. in astronomy at the University of Arizona.

Robert Irion is a freelance science journalist in Santa Cruz, California. He studied earth and planetary sciences at the Massachusetts Institute of Technology and science communication at the University of California, Santa Cruz, where he now teaches a course in writing scientific features for magazines. He is a contributing editor at *Astronomy* and a contributing correspondent at *Science*. His articles on astronomy, earth sciences, and physics have also appeared in *New Scientist*, *Highlights for Children*, and other publications.

Index

Credits

(Key: page numbers in boldface; *(t)* = top, *(b)* = bottom, *(c)* = center, *(l)* = left, *(r)* = right)

All illustrations, with the exception of the figure on page 121, copyright Wood Ronsaville Harlin, Inc.

Front jacket NGC 2997, © Anglo-Australian Observatory, photograph by David Malin; **Back jacket** Hale-Bopp at your back door, Glenn F. Knarr/The News-Item, Shamokin, Pa.; **iii** August 1999 solar eclipse as seen from Neunkirchen, Austria, © Reuters/Heinz-Peter Bader/Archive Photos; **iv-v** Crescent Earth, NASA; **1** stars of the Milky Way, © John Sanford; **6-7** Delicate Arch, Arches National Park, Utah, David Nunuk/Science Photo Library; **15** total solar eclipse, Steve Albers, Dennis DiCicco, Gary Emerson. **16** *(t)* total solar eclipse, Steve Albers, Dennis DiCicco, Gary Emerson; *(c)* annular eclipse, Olivier Staiger; *(b)* lunar eclipse, © Jerry Schad; **19** *(l)* Galileo's notebook, photograph by Robert Emmett Bright, used with permission of Time-Life Books; *(r)* Jupiter and its Galilean satellites, ©1978 Neil deGrasse Tyson; **21** ocean waves, Digital Vision Ltd.; **32** *(b)* water swirling in tub, Koav Levy/PNI; *(inset)* child in tub, Tom Prettyman; **33** *(t)* Hurricane Linda, NOAA; *(b)* and *(inset)* Jupiter's Great Red Spot, NASA/JPL; **36** NGC 2997, © Anglo-Australian Observatory, photograph by David Malin; **40** *(bl)* Abell 2218, NASA/HST; *(br)* B1938+666, L. J. King, NASA/HST; **42** *(t)* Earth photo by NASA/JPL, image processing by W. Reid Thompson; *(c)* Io, NASA/JPL; **43** *(l)* Moon, near side, NASA/JPL; *(r)* Moon, far side, NASA; **45** *(t)* Antennae galaxy, NASA/HST; *(c)* creating tidal tails, Chris Mihos and Sean Maxwell, Case Western Reserve University; *(b)* a galactic bull's-eye, Chris Mihos and Sean Maxwell, Case Western Reserve University; **46** Cartwheel galaxy, NASA/HST; **47** *(tl)* NGC 3509, John Hibbard, NRAO; *(cl)* NGC 1531/2, William Keel, University of Alabama; *(bl)* NGC 4676, John Hibbard, NRAO; *(c)* Polar Ring Galaxy, NGC 4650A, ©Anglo-Australian Observatory, photograph by David Malin; *(tr)* UGC 2320, John Hibbard, NRAO; *(cr)* MCG-3-25-19, William Keel, University of Alabama; *(br)* NGC 2207, The Ohio State University Bright Spiral Galaxy Survey, J. Frogel, R. Pogge, A. Quillen, K. Sellgren, and G. Tiede; **49** Comet West, © Peter Staettmayer; **52** Saturn and its moons and rings, NASA/JPL; **53** *(t)* rings of Saturn, NASA/JPL; **55** Tunguska region impact site, public domain; **58** *(t)* chain of craters on Callisto, NASA/JPL; *(b)* string of pearls, NASA/HST; **59** Jupiter, NASA/HST; **60-61** Comet Hale-Bopp, Brian McLeod; **63** *(l)* Halley's Comet in the Bayeux tapestry, by special permission of the City of Bayeux; *(r)* coin with head of Caesar, Bayerisches National Museum, Munich, Basserman-Jordan Collection; **64** Comet Hale-Bopp, Juan Carlos Casado; **79** Earth and Moon, NASA/JPL; **80** deep-field, NASA/HST; **83** Large Magellanic Cloud, NASA/HST; **85** *(tl)* Eta Carinae, © Anglo-Australian Observatory, photograph by David Malin; *(bl)* Flame Nebula, 2MASS Collaboration, University of Massachusetts, IPAC; *(c)* R Aquarii © Anglo-Australian Observatory, photograph by David Malin; *(tr)* NGC 7635, © Anglo-Australian Observatory, photograph by David Malin; *(br)* NGC 6164-65, © Anglo-Australian Observatory, photograph by David Malin; **90-91** Cygnus Loop, NASA/HST; **95** *(tl)* Herbig-Haro 30, Chris Burrows (STScI), the WFPC2 Science Team, and NASA; *(tc)* HK Tauri, Karl Stapelfeldt (JPL) and colleagues, and NASA; **101** *(t)* Crab Nebula, NASA/HST; *(b)* pulsar in Crab Nebula, N.A. Sharp/NOAO/AURA/NSF; **108-109** SN1987A, © Anglo-Australian Observatory, photograph by David Malin; **113** Super Kamiokande detector, photograph courtesy of Henry Sobel, University of California, Irvine; **116** Meteor Crater, David Roddy; **119** *(t)* a sandy foot, © Brian Bailey/Index Stock Imagery/PNI; *(cr)* Sun's granular surface, NASA; *(b)* Orion Nebula, NOAO; **122** Cassiopeia A, *(tr)* (x-ray), NASA, SAO, Chandra X-ray Center; *(bl)* (optical), R. Fesen, Dartmouth College; *(bc)* (infrared), ISO; *(br)* (radio), VLA/NRAO; **124** Venus, NASA/JPL; **129** Whirlpool galaxy, NRAO; **135** *(cl)* low-pressure sodium lights, Philips Lighting Company; *(bl)* Io, NASA/JPL; **140** *(tr)* graph of star's rotation, Debra Fischer, San Francisco State University; **143** Io, NASA/JPL; **144** solar eruptions, NASA; **147** Earth aurora, David Miller; **152** PG 0052+251, NASA/HST; **153** *(t)* M87 optical jet, NASA/HST; *(c)* M87 in radio, F. Owen, NRAO; *(b)* M87 optical, © Anglo-Australian Observatory, photograph by David Malin; **155** M84, NASA/HST; **156-157** hot springs at Yellowstone National Park, Christy Rothermund and Robert Ramaley, University of Nebraska; **159** Earth view from space, NOAA; **160** *(t)* tardigrade, R.O. Schuster, University of California, Davis; *(l)* tube worms, videograb ROV ROPOS, Kim Juniper (UQAM) and Chuck

Fisher (PSU); *(c)* black smoker, John R. Delaney, University of Washington; *(r)* cyanobacteria, Richard B. Hoover and Greg Jerman (NASA/MSFC), and Sabit Abyzof, Russian Academy of Sciences; *(b)* Moorella obsidium, Christy Rothermund and Robert Ramaley, University of Nebraska; **164** *(l)* NGC 4414, NASA/HST; **165** *(l)* false-color view of Europa, NASA/JPL; *(cl)* cells, PhotoDisc, Inc.; *(cr)* color-enhanced MRI scan of the human brain, Tony Stone Images; *(r)* radio telescope, NRAO; **166** *(l)* Europa's surface, NASA/JPL; *(c)* Callisto's surface, NASA/JPL; *(r)* Martian surface, NASA/JPL; **169** *(t)* Texas meteorite NASA/JSC; *(b)* Martian meteorite, NASA; **171** *(l)* Pioneer plaque, NASA; *(r)* Voyager phonograph record, NASA/JPL; **177** *(l)* COBE sky map, NASA/COBE; *(r)* galaxy clusters, ©1989 SAO; **180** Tevatron, Fermilab Visual Media Services; **181** Heavy Ion Collider, Brookhaven National Laboratory; **182** particle collision, CERN; **186** *(bl, tr, br)* illustrations adapted from *The Elegant Universe: Superstrings, Hidden Dimensions, and the Quest for the Ultimate Theory* by Brian R. Greene, by permission of W.W. Norton & Company, © 1999 Brian R. Greene; **189** *(l, r)* gamma-ray bursts, NASA/HST; **195** *(t)* CL0024+1654, W.N. Colley, J.A. Tyson, E.L. Turner, NASA/HST; *(l)* computer ray-trace simulation, J.A. Tyson, I. Dell'Antonio, and G. Kochanski, Lucent Technologies; *(br)* simulation with too much mass, J.A. Tyson, Lucent Technologies; **196** *(t)* computer simulations, Michael Gross, Patrik Jonsson, and Joel Primack, University of California, Santa Cruz; *(b)* galaxy clusters, ©1989 SAO; **207** *(440 B.C.)* Democritus, Library of Congress; *(c. 260 B.C.)* cosmology of Aristarchus of Samos, from *Dictionary of Scientific Biography*, Volume 1, Charles Scribner's Sons, 1970; *(A.D. 140)* Claudius Ptolemy, Library of Congress; *(1543)* Nicolaus Copernicus, Library of Congress; *(1572)* cover of *De nova stella* by Tycho Brahe, Library of Congress; *(1609)* Johannes Kepler, Library of Congress; *(1610)* Galileo Galilei, Library of Congress; *(1655)* drawing of Saturn by Christian Huygens, from *Popular Astronomy* by Camille Flammarion, D. Appleton & Company, 1925; *(1687)* Sir Isaac Newton, Library of Congress; *(1820)* Hans Christian Oersted, Royal Danish Embassy, Washington, D.C.; **208** *(1831)* Michael Faraday, © Royal Astronomical Society Library; *(1842)* Christian Doppler, Picture Archives of the Austrian National Library; *(1843)* James Prescott Joule, adapted from *James Prescott Joule and the Concept of Energy* as originally published by Watson Publishing International, with permission of the author, Henry John Steffens; *(1858)* Henry Draper photo of stellar spectra, Harvard College Observatory; *(1873)* James Clerk Maxwell, by permission of the President and Council of the Royal Society; *(1887)* Albert Michelson, © The Nobel Foundation; Edward Morley, National Academy of Sciences; *(1900)* Max Planck, © The Nobel Foundation; *(1901)* Guglielmo Marconi, Smithsonian Institution; *(1905)* Albert Einstein, © The Nobel Foundation; *(1909)* Sir Ernest Rutherford, © The Nobel Foundation; **209** *(1912)* Henrietta Swan Leavitt, Harvard College Observatory; *(1913)* Niels Bohr, © The Nobel Foundation; *(1916)* Schwarzschild black hole, Wood Ronsaville Harlin, Inc.; *(1919)* Sir Arthur Eddington, © Royal Astronomical Society Library; *(1925)* detail of periodic table, Wood Ronsaville Harlin, Inc.; *(1927)* Werner Heisenberg, © The Nobel Foundation; *(1929)* Edwin Hubble, National Academy of Sciences; *(1936)* Fritz Zwicky, California Institute of Technology; *(1948)* Ralph Alpher, Ralph Alpher; George Gamow, University of Colorado, Boulder; *(1950)* Fred Whipple, Charles L. Hanson, Jr.; **210** *(1961)* Drake equation; *(1964)* quark, Wood Ronsaville Harlin, Inc.; *(1965)* Arno Penzias and Robert Wilson, © The Nobel Foundation; *(1967)* Jocelyn Bell Burnell, © The Open University; Anthony Hewish, © The Nobel Foundation; *(1974)* Stephen Hawking, Nigel Luckhurst; *(1981)* Alan Guth, Donna Coveney/Massachusetts Institute of Technology; *(1987)* SN1987A, © Anglo-Australian Observatory, photograph by David Malin; *(1990)* Voyager phonograph record, NASA/JPL; *(1993)* Jupiter, NASA/HST; *(1995)* Comet Hale-Bopp, Juan Carlos Casado; **213** photographs of Neil Tyson and Charles Liu, © Harry Heleotis; photograph of Robert Irion, Ron Jones Photography.

Note on the Colorization of Astronomical Images in *One Universe*

Images of astronomical objects in this book are colored according to a palette that represents bands of light outside our familiar visible spectrum. This palette, devised by the Department of Astrophysics at the American Museum of Natural History, provides a color scheme that allows the observer to readily identify the spectral band in which the image was produced. The color palette is displayed in the illustration on page 130. All color photographs taken in the visible spectrum in this book appear in true color, i.e. as the human eye would see the pictured objects directly.

Managing editor	Stephen Mautner
Project editor	Roberta Conlan
Designer	Francesca Moghari
Production manager	Dorothy Lewis
Photo research	Christine Hauser
Illustration	Rob Wood, Matthew Frey Wood Ronsaville Harlin, Inc. Annapolis, Maryland
Printer	Chroma Graphics, Inc., Largo, Maryland

Library of Congress Cataloging-in-Publication Data

Tyson, Neil deGrasse.
   One Universe : at home in the cosmos / Neil deGrasse Tyson, Charles Liu, Robert Irion.
     p. cm.
   Includes bibliographical references and index.
   ISBN 0-309-06488-0 (alk. paper)
   1. Cosmology. I. Liu, Charles, 1968- II. Irion, Robert. III. Title.

QB981 .T97 2000
523.1—dc21

99-053981

The authors extend special thanks to Maron Waxman and her staff at the American Museum of Natural History for their image, content, and editorial guidance on the final manuscript.

Any opinions, findings, conclusions, or recommendations expressed in this volume are those of the authors and do not necessarily reflect the views of the National Academy of Sciences or its affiliated institutions.